From Betamax to Blockbuster

Inside Technology

edited by Wiebe E. Bijker, W. Bernard Carlson, and Trevor Pinch

A list of books in the series appears at the back of the book.

From Betamax to Blockbuster

Video Stores and the Invention of Movies on Video

Joshua M. Greenberg

The MIT Press
Cambridge, Massachusetts
London, England

For information about special quantity discounts, please e-mail special_sales@mitpress.mit.edu.

This book was set in Stone Sans and Stone Serif by SPi Publisher Services, Pondicherry, India.

Printed and bound in the United States of America.

Library of Congress Cataloging-in-Publication Data

Greenberg, Joshua.
From betamax to blockbuster : video stores and the invention of movies on video / by Josh Greenberg.
     p.   cm.
Includes bibliographical references and index.
ISBN 978-0-262-07290-8 (hardcover : alk. paper)
1. Videocassette recorders. 2. Video recordings industry—History. I. Title.

TK6655.V5G74   2007
384.55'8—dc22

2007018942

10 9 8 7 6 5 4 3 2 1

Contents

Acknowledgments

This book would not have been possible without the help of various members of the video industry, both past and present. Too numerous to name here, they are the heroes of this story; from New York to California, Cleveland to Florida, I am in debt to these men and women for their generosity in both time and honesty. Thanks in particular to Mark Wielage, Rod Woodcock, Ray Glasser, and Rich Nathanson for opening their video and print archives to me, as well as for hours of conversation in their homes. I can only hope that I have done their stories justice.

Many preliminary interviews were done at trade shows, made substantially easier by press credentials extended by both the Consumer Electronics Show and the Adult Entertainment Expo. Additionally, representatives of the Consumer Electronics Association, the Custom Electronic Design and Installation Association, and Adult Video News helped to flesh out both the history of their respective organizations and of the video industry as a whole. Of particular note, Brad Hackley, Mark Fisher, and Carrie Dieterich of the Video Software Dealer's Association sat down with me around the VSDA conference table, and then put me in touch with many of the key figures in the history of video retail.

For access to trade journals and other primary materials, I found the collections of the New York Public Library, the Los Angeles Public Library, the UCLA Film Library, and the Library of Congress simply invaluable. I'm indebted to Ron Roache at the Library of Congress for setting up carts of materials in advance, saving me hours (if not days) of filling out little pink slips, as well as to Alex Magoun for a fascinating day spent perusing a spread of artifacts and documents in the RCA archives at the Sarnoff Corporation and to Don Rosenberg at *Video Store* magazine for letting me rummage through back issues at the magazine's offices.

One untraditional aspect of my research was the use of a website to collect memories and stories from people on both sides of the video store counter (see the appendix). My eyes were first opened to this method during an Exploring and Collecting History Online weekend seminar, sponsored by the Center for History and New Media at George Mason University, and I'm particularly grateful to Dan Cohen and Jim Sparrow for guiding me through the process of designing and deploying videostoreproject.com. As of this writing, the site has received well over a thousand stories, thanks in large part to a kind post by "Moriarty" on the film gossip site *Ain't It Cool News*, which prompted a veritable tidal wave of respondents.

As the place where I first formulated this project, Cornell's Department of Science and Technology Studies was a true intellectual home, where the community model of graduate education flourished (though some faculty merit particular thanks: Trevor Pinch, Phoebe Sengers, Michael Lynch, and Michael Dennis). Virtually every one of my fellow graduate students contributed in some form or another to this dissertation: however, special thanks are due to Adelheid Voskuhl, who has been a steady and thoughtful presence since our first days in Ithaca. Most importantly, two mentors bear singling out: Bruce Lewenstein and Ron Kline both taught me the importance of looking beyond traditional categories of expertise, whether in the context of the public understanding of science or in the social context of technology use, and I'm richer for their guidance and friendship.

After leaving Ithaca, the Center for History and New Media at George Mason University offered a unique opportunity to indulge both scholarly and geeky tendencies—it was a luxury to spend my days working alongside colleagues like Roy Rosenzweig, Dan Cohen, Tom Scheinfeldt, Kelly Shrum, and Sharon Leon. I also was lucky to stumble into a writing group soon after arriving in D.C., and Alex Russo and Elena Razlogova helped immensely as I retooled the dissertation into a book. Beyond the immediate area, an extended network of colleagues and friends have read, heard, and argued with various pieces of this project, and thanks are due to Tarleton Gillespie, Fred Turner, Pablo Boczowski, Amber Watts, Shay David, Jonathan Coopersmith, Bart Simon, Greg Downey, Jofish Kaye, Max Dawson, and especially Randy Cohen, on whom I can always count for a 2 a.m. chat when I'm stuck.

My immediate family has been a source of constant support: Barry Greenberg, who took me to work at United Wholesalers and introduced me to the world of wholesale distribution; Kay Greenberg, who taught me to love knowledge; Steve McConville, who offered a warm place in Florida to regroup every winter; and my sister Alisha, who keeps me grounded and focused on the task at hand. To my extended family of Pitlers, Greenbergs, McConvilles, and Bilenkers, thank you for your support and interest—I promise that you'll never have to hear about the early video industry at a family event ever again.

Finally, Jenny Bilenker, who married me knowing the full extent of my capabilities for distraction, near-obsessive tunnel vision, and erratic sleeping: she made room for me in her heart and carved out an office for me in her apartment (no mean feat in New York City). Thank you.

Introduction

You probably remember your first video store.

If you're older than a certain age, it was a smallish space jammed with shelves, with movie posters on the walls and a membership club you had to join. Whether a storefront in a suburban strip mall or a below-ground space on a city street corner, you probably knew the owner by name, and he or she likely knew your taste in movies. Odds are that the shelves were lined with empty videocassette boxes that would be exchanged at the counter for tapes in dark plastic cases, and you might have been able to rent a briefcase-like video player or video camera as well. There was surely a back room of some kind into which children weren't allowed, and the color scheme was most definitely *not* yellow and blue.

They probably stocked Betamax.

It's likely that you don't go to that store today, if it's even still in business. You might rent movies from a massive franchise store, or you might get them by mail, fresh off your online queue; either way, you're probably renting DVDs, not videocassettes, and the guy behind the counter has been replaced by data-aggregating recommendation algorithms. Maybe you don't deal with tangible media at all, watching movies-on-demand via your cable company or downloading them (either legally or not) via the Internet. You may own a collection of movies, stacked either physically alongside your television set or virtually in a program like iTunes. You might view them in your living room on a television set, in your office on a computer, or on an mp3 player while commuting to work.

The humble video store and its enabling technology, the videocassette recorder (VCR), laid the foundation for this world in which movies are tangible consumer goods. In order to understand how consumers came to take for granted the ability not just to possess movies but also to watch them when, where, and however they chose, we need to look back to the invention of the home video industry itself. In 1976, when Sony framed the first Betamax solely as a time-shifting device, none of this was supposed to happen. Initially, the rentals and sales of prerecorded movies on videocassette weren't encouraged by either movie studios or electronics manufacturers, but within just thirteen years the motion picture industry was making more money from video than from theatrical exhibition.[1] This book tells the story of how it all got started.

The Canonical History of the VCR

The history of the VCR as a technology is surprisingly tricky to pin down, in no small part because it has lived at least three distinct and often overlapping lives. First, the machine can be a time-shifting device, an extension of broadcast television technology. Secondly, users have combined VCRs with video cameras to author their own movies, most traditionally of family events.[2] Finally, the VCR can be used to view movies and other content on prerecorded tapes, creating a version of the movie theater in the home. These three "meanings" of the VCR are by no means mutually exclusive, coexisting to varying degrees in households across the United States, and VCR owners shift seamlessly between them. A comprehensive history of the VCR would examine each of these uses, but that would be a very different book than the one you're holding in your hands; the story here is of one particular use, engaging other practices only when they directly touch on the rental and sale of prerecorded movies on videocassette.[3]

The general history of movies and the VCR has been told enough times by enough people that a canonical version has emerged, one of the sort that a business professor might throw into a lecture as a case history. The general contours of this story can be summed up in two parables and a bit of folk wisdom: *Universal v. Sony* (the "Betamax Case"), VHS and the death of Betamax, and pornography.

First, the story of the VCR has often been couched as the triumph of David over Goliath, with Hollywood playing the part of the giant and home users wielding the metaphorical slingshot. When first introduced by Sony, the Betamax was advertised as a machine that would allow users to assert control of their time, liberated from the fixed program schedules set by television executives—as one Sony ad proclaimed, "Now you don't have to miss *Kojak* because you're watching *Columbo* (or vice versa)!"

This call for freedom was particularly ill received by Universal Pictures, which happened to produce both *Kojak* and *Columbo* and whose president, Sidney Sheinberg, spearheaded a lawsuit against Sony to stop Betamax sales entirely, arguing that any such off-air recording constituted an infringement on his company's intellectual property.[4] The ensuing court battle, which began in 1976 and lasted until the Supreme Court handed down a decision in 1984, has been recounted many times in many places, with a particularly detailed account in James Lardner's essential book *Fast Forward: Hollywood, the Japanese, and the Onslaught of the VCR*.[5]

Notably, the lawsuit was filed not just against Sony, but also against an individual Betamax owner, William Griffiths, as well as the store where he'd bought the machine. Though Griffiths was in fact a willing accomplice (he was a client of the same firm that was representing Universal, and had agreed to be sued under the condition that no damages would be sought), many other VCR owners saw the court case as an attack on their own personal liberties, a frame soon taken up by the popular culture constellation of magazines and newspapers. By the time the U.S. Supreme Court decided by a five-to-four margin that home taping did *not* constitute copyright infringement, setting in place a new understanding of the fair use of intellectual property,[6] public opinion was decisively on the side of Sony and the ally it had played a part in forming, the Home Recording Rights Coalition, which was seen as fighting for the rights of citizens everywhere against the media behemoths. At this point, even Hollywood studios themselves were ambivalent about the lawsuit, and Lardner's book offers a cautionary tale of how the Hollywood studios clung to traditional views of the motion picture until they were finally swept away in the unstoppable tidal wave of the emerging video retail/rental market.

The second canonical moment in the VCR's history concerns the contest between two competing technical standards, VHS and Betamax. Though Sony's Betamax was first into the market, it quickly found a strong competitor in the Japan Victor Company's VHS (for "Video Home System"), which was soon adopted by Matshushita.[7] From a technical standpoint, the main difference between the two formats was image quality versus tape length; Betamax cassettes recorded sharper and clearer images and sound, but could only hold an hour of video as opposed to VHS's two-hour recording time. Though Sony soon responded with a quickly constructed "Betastack" tape changer, and then a two-hour cassette, its North American rival RCA had already gone even further by introducing a half-speed VHS recorder that allowed up to four hours of taping (RCA executives had chosen to ally with the VHS standard under the condition that Matsushita's engineers could figure out how to create a machine capable of recording an entire American football game).[8]

The competition between VHS and Betamax is often framed as a parable about the success of mediocrity over higher standards, not unlike the way many describe the fate of the Apple computer relative to the pervasive success of Microsoft's successful DOS and Windows operating systems. The actual causes of this success seem less important than the larger moral: one author suggests that the fault simply lay with Sony's lack of foresight in marketing a system with shorter tape life, while another sees a lesson about the public's preference for frugality over excellence, and still a third points to RCA's framing of the VCR as a technology for the everyman, as opposed to the high-tech Betamax.[9]

Regardless of the cause, VHS had a larger market share than Betamax by the end of 1978, but Cusumano, Mylonadis, and Rosenbloom argue in a 1992 *Business History Review* article that Betamax might have maintained a stable portion of the market (again, much as Apple has managed a stable, if small, percentage of the computer market) were it not for the "network externality" of prerecorded cassette tapes in the early 1980s. "RCA had well-developed ideas about the consumer market for recorded video programming," they write, but "Sony also proved less effective than Matsushita in supplying equipment for duplication of tapes in the Beta format." As video sale and rental became more important, their argument goes, "[t]he greater abundance of VHS program material gave buyers greater incentive to choose VHS players, which then led tape distributors to stock more VHS tapes, in a reinforcing pattern."[10] Sales lagged, and by 1985, Sony had begun to scale back production of Betamax recorders (though the format survived among professionals and semiprofessionals for quite some time afterwards).[11]

Finally, there's the question of pornography. An awkward consequence of writing a book on the history of video stores is that whenever it comes up in conversation, one of the first questions is invariably "So, you're looking at porn, right?" Among scholars of technology, Jonathan Coopersmith in particular has remarked on the role that pornography seems to play in the early history of individual media technologies, and virtually every history of the VCR nods and winks toward the link between adult film and the early days of prerecorded video.[12]

At the same time as video stores began to pop up around the country, a parallel trend of moral conservatism was gaining prominence. With the founding of the Moral Majority in 1979, issues of morality and decency were thrust to the foreground of American culture, and video stores were no exception. Unlike adult theaters, video stores opened in "good" parts of town, and some citizens were upset to find that pornography could be acquired in the same store where their kids rented *Sleeping Beauty*. In some cities, it became common to see church members picketing outside the local video store, upset at the incursion of immorality into their neighborhood.[13] Meanwhile, the Moral Majority found an unlikely anti-pornography ally in the feminist movement, which linked adult videos to violence against women.[14] Though the mainstream video industry made a deliberate effort to move beyond the back room toward a more family-friendly image, the stigma of pornography's early importance to the technology (and the industry) never quite washed off.

While these canonical tropes in the VCR's history are core parts of its story, they are not my primary focus. Instead, this book aims to flesh out a part of this history that has been generally taken for granted: the basic transformation of the VCR from a machine that records television to a machine that plays prerecorded cassettes, generally movies. This was for the most part not a physical transformation, since the technical aspects of VCRs facilitated both uses. Instead, it was a transformation in meaning, involving the construction of new (though not necessarily contradictory) knowledge about what a VCR was good for, as well as the more literal construction of a network of social institutions that would support this new meaning in the consumer marketplace. In short, this is the story about how the VCR was remade into a medium for movies, and who was responsible for doing so.

Mediators and Consumer Technology

In his play *Rosencranz and Guildenstern are Dead*, Tom Stoppard focuses on two of the seemingly least-important characters of Shakespeare's *Hamlet*, weaving a narrative that takes place alongside that of the melancholy Dane. Throughout Shakespeare's version, Rosencranz and Guildenstern are relatively inconsequential characters who act as the instruments of others; their chief role in the play is to convey Hamlet's death sentence (which ultimately becomes their own). They are minor characters precisely because they exist in between the major characters, with no real identity or purpose of their own. In his riff on the play, however, Stoppard puts Rosencranz and Guildenstern at the center of the story, relegating Hamlet, Claudius, and the rest of the royal family to the periphery. By focusing on the action in the passageways and rooms just beyond *Hamlet*'s edges, Stoppard offers an alternate perspective that foregrounds the importance of his protagonists' mediating function and serves as a fatalistic meditation on their powerlessness in the face of those more "important" characters between whose interests they are trapped.

There is a useful lesson here for the history of technology. Having moved away from a view of the history of technology that is about the production of artifacts to one focusing on the production of knowledge about the nature and use of those artifacts (essentially studying *ideas* about things rather than *things* per se), the question of who should figure as protagonists becomes problematic. Many sociologists of technology respond that we should focus on "relevant social groups," those groups of social actors (bound together by a common identity) who have the skills and credibility to define a technology's cultural meanings. Naturally, the most readily apparent social groups are those, like Claudius or Hamlet, who speak the loudest (and who are the most socially prominent). In most accounts of the history of the VCR, these most prominent actors have been the Hollywood studios, the technology manufacturers, the government, and the users themselves. Rather than cover ground that has been well trod by others, however, I'll spend the following pages charting a parallel history, one which takes place in the less-explored spaces *between* the media corporations, technology manufacturers, and lawmakers who are the traditional protagonists of the VCR's history.

Such "in-between" spaces can be quite important, yet easily overlooked. A consumer technology may pass through many hands between its manufacturer and its user, and in the case of movies on video, movies weren't the only things being mediated. Just as a movie director can't simply deposit her completed film in the mind of an audience member, a VCR manufacturer can't directly place the machine in the hands of a user; in both cases, layers of mediation help to package, distribute, and sell the product. The more successful the movie or machine, the more important and necessary the role played by this distribution network (and the more complex the network itself). If we are to take the perspective that the history of a technology is essentially the history of knowledge production *about* that technology, then it seems natural to look not merely at its producers and consumers, but at the spaces in between them through which such knowledge is mediated. Though these levels of mediation might seem to be transparent channels between producer and consumer, the mediators populating them play a surprisingly active role in the production and maintenance of knowledge about their goods, a role that has gone relatively unexplored. Such mediators can be understood much like media in terms of their "in-betweenness," though the term "mediator" carries a more deliberate connotation of agency than "medium" (implying a greater sense of transparency between sender and receiver).[15]

In the early days of movies on video in the United States, corporations were continually playing catch-up with the small business owners and enthusiasts who were creating the video industry and who mediated home users' access to the new technology. This isn't a story of mediators in a preexisting sociotechnical network—these distributors and mom and pop video retailers, often characterized as isolated homesteaders on the wild frontier of home video, were in fact responsible for creating new meanings for and uses of preexisting technology within their own distribution networks. Film studios, manufacturers, and home users of the time were mainly focused on the VCR as a time-shifting technology, and to varying degrees even resisted alternate technological frames. Video distributors and retailers enrolled these other social groups, setting themselves up as what Latour calls "obligatory passage points" in this new system of artifacts and practices.[16] While a few of these mediators are mentioned (and, in the case of one chapter in Wasser's book, even foregrounded), for the most part their stories are as of yet untold—or worse, seen as a transitional phase before the real story, the consolidation of the corporate franchises that ultimately took over the industry.[17]

I want to argue that the invention of movies on video took place in the space populated by video retailers and distributors, in-between audience members and the video hardware and software manufacturers. Ruth Schwartz Cowan describes this as the "consumption junction," explaining that we can make sense of the disparate network of social actors involved with a consumer technology by looking at this broader landscape from the perspective of consumers, on whose shoulders a technology's success or failure lies. According to Cowan, one can understand this history by examining the options made available to consumers and how those consumers ultimately spend their money (in effect choosing a technology).[18]

While useful in explaining technological successes and failures, a consumer's-eye perspective is inherently limited, simply because it doesn't help to explain how or why a given set of options was presented to consumers. To tackle that question, we must work our way up the distribution network from the consumers, putting mediators like distributors and retailers at the center of the story and accounting for the development of home video from their perspective. Decisions about whether a videotape should be sold or rented, where it should be procured, and how movies on video should be understood were the product of negotiations between these different actors, played out not in the board room or the courtroom but at the various layers of distribution and retail that ultimately shaped the consumption junction where consumers spent their money.[19] Moreover, these mediators handled more than just the goods themselves, and their mediation of knowledge about the VCR and movies on video was arguably as important as the shelves they stocked.

One consequence of a consumption junction-oriented focus on mediators is a certain narrowing of scope. For example, the VCR's material nature was the product of many negotiations between engineers, marketers, and corporate executives, and Graham documents these sorts of processes wonderfully in her study of the RCA videodisc player.[20] My choice of the mediation space as the place from which to study the history of movies on video, however, means that I only take the actions of manufacturers into account insofar as they involved direct interaction with mediators, thus influencing the consumption junction. Negotiations about the material nature of the VCR lay beyond the reach of most distributors and retailers—VCRs were (quite literally) solid black boxes that arrived on their loading docks or were already in their customers' homes. For the most part, this means that my story is less about the materiality of the VCR and more about the building of new knowledge about that same physical machine (though at moments this new knowledge does ripple back into a material change in the artifacts themselves, as well as the physical spaces of the consumption junction). Moreover, I've focused my account entirely within North America both for reasons of practicality and space; the history of video in Great Britain, for example, would be a book unto itself.[21]

The VCR as a Medium (or, A Brief Romp through Media Theory)

Though whole academic and professional industries are dedicated to its study, the general category of media and communication has been surprisingly undertheorized by historians and sociologists of technology. Since this book charts the refashioning of the VCR into a medium for prerecorded movies, it seems worth spending a few of them getting a solid foundation under our feet.

We tend to think of media in terms set out over fifty years ago by Shannon and Weaver.[22] In their canonical model, a channel sits between a sender and a receiver, carrying information. This channel is not viewed primarily in terms of its in-betweenness, but rather in terms of the information that it carries from one side to the other—it, is the very act of carrying information, in other words, that makes something a medium. According to this model, the system of sender, channel, and receiver is functionally understood in terms of the information that it mediates, described and analyzed in the context of use rather than through any intrinsic characteristics.

This leads to an interesting point with regard to the nature of media—a particular sort of use distinguishes media from other technologies, nothing more. There is no inherent reason that we should treat somebody watching a television set differently from somebody watching a hammer or a fork, but we do, because in the former case we come to our observation with the a priori knowledge that the television brings information to the user from some distant sender.[23] Such information goes by many names ("content" or "signified," for example, depending on your scholarly genealogy) but henceforth I'll use the term "message," in no small part because the historical baggage that the term carries helps to frame medium and message as two sides of the same coin, yin and yang comprising a whole.

The question of how to study media and their associated messages has occupied many scholars, particularly in the era of mass communication. Through the twentieth century, the dominant approach in studies of propaganda and the developing field of Communication focused on the information itself, as scholars like Harold Lasswell and Paul Lazarsfeld studied the ways in which media messages impacted society.[24] While most of their intellectual heirs have moved toward a more active conceptualization of audiences as favoring particular messages based on their "uses and gratifications," the legacy of this message-oriented stance still persists in much work on persuasion and content analysis.[25] Taking this trend to its logical conclusion, other communication theorists have followed the general rise of social constructionist scholarship, using ethnographic, sociological, or cultural studies methods to understand the relationship between audiences and the messages they interpret, as well as the construction of the audience itself.[26]

While sophisticated in their analyses of messages, mass media scholars have tended to be less engaged with the material and technological nature of the media themselves. For the most part, these aspects of media fall into the domain of historians and sociologists of technology, and more recent work in this vein has fleshed out the development of new sociotechnical systems, foregrounding the social negotiations involved in the construction of technologies such as radio,[27] the telegraph,[28] the telephone,[29] television,[30] sound recording,[31] and the recent development of the Internet.[32] This body of work is rich in empirical explorations of the roles of various social groups (particularly users) in the construction of these new technologies, but tends to gloss over the actual information they mediate. Meanwhile, many cultural historians *have* explored the content of media technologies, but usually at the expense of a nuanced understanding of the social dynamics of the technologies' construction[33]—rare is the historian of technology who manages equal sensitivity to issues of both medium and message.[34]

Perhaps the closest overlap between the fields of media studies and technology studies came in the 1960s, when Marshall McLuhan turned traditional media theory on its head with his famous dictum, "The medium is the message."[35] Countering the "hypodermic needle" theories of mass communication research, McLuhan's Toronto School movement believed that the content carried by a medium was irrelevant to its impact (what he saw as the *real* "message") on society. Instead, the argument went, the inherent artifactual nature of a communication technology shapes its impact, a technological determinism that became central to the emerging subfield of media ecology. While McLuhan was generally optimistic about the possible impact of emerging technologies, his ideas were picked up by Neil Postman, who fused them with the Frankfurt School's general pessimism toward mass culture, arguing that electronic media erode the rational, literate culture of the Enlightenment.[36] More recently, some scholars in this tradition have turned away from Postman's normative stance and begun to fuse the media ecological perspective with other approaches such as Goffman's situational interactionism[37] or a "softer" form of determinism.[38]

While most research on communication technologies has focused on the relative importance of either the medium or the message, I argue that their separation into distinct categories is a social construction rather than a necessary schism. In order to use a medium skillfully, a user needs to be able to take two steps: first, she must envision the technology as sitting in-between herself and some sender rather than simply existing on its own; and second, she must be able to distill meaningful information from the experience of using the technology. Thus, *any* technology can be constructed as a communication technology—all it takes is a consensus among its users that information is being mediated. This mediated information is a part of the set of meanings and practices that Bijker calls a "technological frame," shared among members of a given social group and informing their use of the technology.[39] In fact, I would argue that the understood presence of a message being mediated is the very aspect of a technological frame that distinguishes a given technology as a proper *communication* technology.[40]

Such a message is usually understood as ontologically distinct, inhabiting the vessel of the medium much as a Cartesian soul resides in a physical body. This vision of a free-floating text seems particularly persuasive given that we can turn something like a radio on and off—with a single action, a seemingly mute artifact springs to life. The fact that we can see a communication technology in both "dead" and "alive" states reinforces the idea that information exists in an immaterial realm, entirely distinct from the medium, which simply carries it.[41]

Consider a television set: it is essentially nothing more than a box with one face that changes colors. In order to meaningfully use it as a television set, a skilled viewer first proceeds under the assumption that the machine is receiving a signal from somewhere else; in other words, the television set is situated between herself and a broadcasting sender.[42] Once that leap is made, it falls to the user to distinguish the meaningful information that is being mediated by the television from its noninformational attributes, and one of the lessons of the social construction of technology is that she must do so without recourse to the innate properties of the artifact itself.

Since there is no innate property that makes its kinetic surface more meaningful than its static surfaces, television use involves a learned ability to distinguish the parts of that box that are meaningful from those that are not. As certain aspects of television as a technology are isolated and labeled as "message," everything else becomes the "medium," that which carries the message. Just as Bruno Latour describes the coproduction of nature and society when a boundary is drawn through human experience, we produce the yin and yang categories of message and medium through the practice of media use.[43] This boundary between medium and message can be remarkably sophisticated; when we see white flecks within the physical boundary of the television screen, for example, we immediately understand them as static, an artifact of the medium rather than a part of the message, even though by all appearances they seem a part of the image on the screen.

As one of my favorite seminar interlocutors might ask, "What's the cash value of this approach?" First, by understanding that an essential aspect of a medium is its in-betweenness, we can begin to look at it from the perspectives of those on either side, exploring the ways in which someone on the receiving end constructs an idea of the sender (and vice versa).[44] The question of how to study media and messages is reframed—changed from the impact of either on users to the ways in which users interpret *both* medium and message. Meanwhile, the precise placement of the boundary between medium and message can be a useful assay to distinguish various social groups (think, for example, of hi-fi enthusiasts, who may perceive their stereo hardware's influence on a sound that seems perfectly pure to a nonenthusiast listener). In this sense, this inherent property of media can be understood as a boundary object, another sort of in-between creature.[45]

Getting back to the subject at hand, however, sensitivity to these questions is essential in telling the history of how a technology becomes a medium in the first place. The consumer VCR was initially conceived as a time-shifting device, a storage technology allowing users to record broadcast television for later playback. It was *not*, however, seen by its early adopters as a medium, because it did not sit between a sender and a receiver in any essential sense—the television set was the medium of broadcast television, and the VCR was initially added to the broadcaster-television-viewer chain not in series, but rather as a peripheral on the viewer's end (see figure I.1). By the late 1980s, however, the VCR had been reconstructed as a medium in its own right, through which motion pictures moved from Hollywood studios to home viewers and for which the television itself was an accessory that facilitated display—these were "movies on video," not "movies on television."[46] These movies on video, meanwhile, didn't look precisely the same as, nor was the experience of watching them exactly similar to, those in the traditional movie theater. In order for movies on video to be considered tantamount to movies in the theater, the differences between the two needed to be explained as consequences of the differing media of film and video (rather than differences in the messages themselves) by the individuals selling them. Thus, the boundary between a movie and the technology mediating it was redrawn in the video era, with consequences that persist to this day.

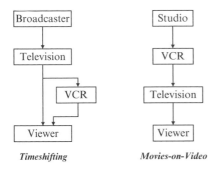

Timeshifting   *Movies-on-Video*

Figure I.1
Two different understandings of the VCR's role in the mediation of information. In the case of time shifting, the VCR was framed as an accessory that would help the viewer manage her end of the communication process, but one that was peripheral to the technological medium of television broadcasting. In the case of movies-on-video, however, the VCR becomes the primary conduit through which the movie moves from studio to viewer, with the television set a necessary addition to display the movie encoded on the cassette tape.

A Brief Outline

This, then, is a book with two arguments: first, that the VCR was refashioned from a time-shifting appendage for broadcast television into a medium in its own right for movies and other prerecorded content; and second, that this refashioning was performed not by producers or consumers, but by those distributors and retailers who occupied the space between them.

This story begins with a look at the practices of early video enthusiasts, who fundamentally understood the videocassette recorder as a technology that extended their television viewing over space and time. From the perspective of the sociology of technology, this is a more traditional case; these enthusiasts were a small, relatively coherent group with clear interests and a shared set of cultural norms. While establishing a baseline of the ways VCRs were used before the advent of video rental, this first chapter also offers a case study of the particular norms and practices of the videophile community, exploring the ways that its members "hacked" both the technical and social nature of their VCRs. The chapter charts the formation of a community of enthusiasts through newsletters, phone networks, and in-person conventions, reflecting on the importance of a few key mediating figures in the formation of an information-sharing network, as well as the ways in which tape-trading and reciprocal off-air recording allowed VCR owners to become part of a larger sociotechnical system. As the VCR became more common-place, however, the enthusiast community began to wane, and was ultimately marginalized by the mass of nonenthusiast users.

Chapter 2 focuses on videocassettes themselves, charting the process by which tangible pieces of consumer technology were redefined as a transpar-ent medium for a far more valuable product, movies. Initially, blank cassettes were accessories for VCRs, no more special than head cleaners or other main-tenance items. It was only after mainstream movies were released on prere-corded videocassettes that the cassettes began to assume an importance of their own. This was far from an easy process, however, and the early domi-nance of public domain and pornographic videos speaks to the reluctance of studios to release their movies in the new format. Those responsible for making this happen were the new video distributors, who mediated between retailers and manufacturers of recorders, tapes, and movies, facilitating a grad-ual movement from a distribution model based on the brown goods world of high-technology appliances to one based on the recording industry.

Once prerecorded videocassettes were available for sale, the next step was the creation of a new consumption junction that would actually bring them to consumers, and chapter 3 begins with a survey of the ways in which mov-ies on video were initially shoehorned into existing retail spaces. From cam-era stores to electronics stores to truck rental stores, retailers from a variety of backgrounds each saw home video through the frame of their own experi-ence, reflecting the interpretive flexibility of the technology in those early days. Over time—thanks in large part to the video distributors, who acted as knowledge brokers and helped shape a shared consensus on how to market and sell movies on videocassette—entrepreneurs developed a wildly success-ful type of store that framed movies on video in terms of the movie theater, unleashing the "gold rush" days of the video industry.

Arguing that the retail space is a physical manifestation of a technological frame, chapter 4 analyzes the spatial layout of the newly established video store. Answers to questions as basic as whether to rent or sell movies led to very clear choices in store design, and simple decisions such as whether cas-settes were kept behind the counter or were "live" (available for customers to handle on their own) reflected a deeper understanding of the nature of the products for sale. At the same time, many stores began to offer redesigned VCRs for rental, making the ability to view a movie in one's home even more ephemeral. As video stores oriented themselves toward offering the temporary experience of a movie (via rental) rather than concrete goods for sale, they began to mimic the visual rhetoric of the movie theater, selling refreshments like candy and popcorn underneath marquee lights and poster displays. While strongly influenced by the movie theater, video stores also reflected the uniquely domestic nature of their technology, and most stores contained two uniquely demarcated sections reflecting basic functions of the domestic space, sex (the adult video "back room") and childrearing (the children's video section).

Looking back toward the consumption junction writ large, chapter 5 explores the social dynamics of the video store, from its role in the broader cultural life of a community to the expertise of the clerk behind the counter (as well as the implications of that expertise outside the bounds of the video store itself). For a time, the video store functioned as a "great good place," and the chapter charts the ways in which many a video store functioned, as one storeowner put it, as "a bar without the alcohol." In many ways, video stores adopted the social functions of the movie theater in American life: though patrons were watching movies in their homes, they were still "going out" to the store beforehand. Storeowners and clerks became figures of authority, but the orientation of the video store toward movies rather than technology meant that their expertise was more filmic than technical. In some cases, video store workers managed to parlay this expertise into a broader expert identity and, in a few rare cases, fame and fortune. For most, however, the local video store was simply a place of friendly camaraderie and conversation.

By the late 1980s, however, the grassroots nature of early video stores was replaced by a more top-down, centralized industry, and chapter 6 charts the ways in which informal information channels and knowledge brokers were replaced by homogenizing trade organizations and franchises. These top-down institutions served as closure mechanisms, establishing and mandating a particular understanding of exactly what movies on video were, and thus how the video store should function. As closure about the nature of movies on video was gradually achieved, one corporation in particular leveraged this newly stable technological frame into a nationwide network of identical consumption junctions. The chapter ends with this relatively stable network of Blockbuster Video storefronts, albeit alongside a small subcurrent of independent video stores in major urban and rural areas.

Finally, the narrative takes a step back to consider the implications of this new medium for preexisting communication technologies. Before video, the fact that movies were meant to be seen in theaters was unquestioned, and ancillary media (like television) were little more than necessary evils. Once home video was established as tantamount to (or at least as acceptable as) the movie theater, the relationship of movies to the technology that mediated them was suddenly thrown in question. The transformations involved with putting movies on videocassette, from the different shapes of the theater and television screens to new possibilities for colorizing black-and-white films, problematized previously stable understandings of what a movie was and should be, necessitating the reevaluation of where exactly the boundary lay between the movie and its medium.

The underlying theme running through both of these arguments, and the book as a whole, is the importance of mediation. On one level, VCRs and movies on video literally moved through a network of mediators in order to get from producers to consumers. A closer examination of these nearly invisible mediators poses challenging questions to social theories of technology, which are too often enamored with those who make and those who use to see all that comes between. At the same time, the new video technology was itself a medium, bringing movies into homes around the country and transforming movies from experiences into commodified goods. The book shows that in both cases, this process of mediation was far from transparent—in fact, it played a central role in the definition of both message and medium . . . software and hardware . . . movies and machines.

1 Videophiles and Betamania: Hacking the VCR

"This is madness."

The first image on the old, hand-shot videotape is a young man named Art Vuolo, who is holding a microphone and wearing a t-shirt that reads "I look better on videotape (really)." Dozens of voices in the background laugh and make wisecracks. Art, barely holding back a laugh himself, tells the crowd to "hold it down."[1]

"It's Fremont, Ohio, 1979. This is probably the most bizarre thing that yours truly has ever witnessed, and in the next couple of minutes, or however long this runs until we are removed from tape or we disintegrate the heads, I'd like to show you what's going on here on this, the fifth of May."

Art pauses for a friendly argument with the cameraman, who has started to pan around the room. "Don't pan around the room until I *start telling people* what's in the room, Ray!" After a brief scuffle, he resumes his introduction. "Let me show you what the heck we've got. Down here . . . we have a variety of machines. It looks like a Highland appliance store!"

The camera pulls back from Art to reveal a hotel meeting room, the central conference table covered with VCRs. There are thirty nine video recorders in the room (mostly Betamax, with a few U-Matic three-quarter-inchers and an RCA VHS or two), not to mention the guest of honor, the brand-new MCA Discovision laserdisc player. The machines are all wired together in various states of recording and playback, the outputs of one plugged into the inputs of the next.

Panning around the room, we see dozens of men (and, notably, one woman) joking around, tending to the machines and generally having a good time. They have brought their recorders, machines that weigh as much as sixty or seventy pounds each, from as far away as Michigan, Missouri, and even New Hampshire. One drove hundreds of miles with his leg in a cast;

another came even though his wife was due to give birth any day. They had come for the second convention of the Video Collectors of Ohio, and they were staying up all night, fueled by cans of Coca-Cola and cups of coffee, talking excitedly about technology, dubbing videotapes from *The Rocky Horror Picture Show* and *Love Story*, and reveling in the company of fellow videophiles.

Film Collectors and Videophiles

In the late 1970s, when the videocassette recorder was a new and expensive consumer technology and video stores were unheard of, many early adopters found themselves infatuated with the machine and what it could do. As one hobbyist wrote, these videophiles were "usually male, 21 to 39 years old, often single (because they spend so much time with their machines they have little time left to be sociable), rarely look the way they sound on the phone, rarely sound the way they write . . . are hospitable, trustworthy, generally reliable, and all have enormous *telephone* bills!"[2] Another enthusiast "considered anyone with more money invested in tape than machines a videophile; another suggested it's anybody who has at least two VCRs."[3] Regardless of their exact definition, videophiles (who defined an identity based on their active and enthusiastic relationship to the new technology of home video) were without doubt a booming subculture.

This hobbyist video culture really began with Sony's introduction of the Betamax in late 1975, but individual movie enthusiasts with home collections had been around for decades beforehand. Early enthusiasts collected 16mm prints of films, and when Eastman Kodak introduced the 8mm film format in the 1960s, film collecting followed.[4] By the mid-1970s major studios like Columbia and Universal had established in-house companies that released condensed "digest" versions of popular motion pictures on 8mm for home use, and publications like *Film Collector's World* and *The Big Reel* were used by collectors to advertise what they had for sale, as well as the prints that they were seeking.[5]

While the introduction of 8mm film lowered the cost of both equipment and film enough to bring movies into the homes of many Americans, in the pre-Betamax days a hardy few had moved into the realm of home video by purchasing professional video equipment and setting up home systems. Perhaps the most famous video hobbyist of these early days was Hugh Hefner

Figure 1.1
Art Vuolo at the 1979 Video Collectors of Ohio Convention. (Video stills courtesy of Ray Glasser).

of *Playboy* fame, who outfitted both Playboy mansions with Ampex open-reel video recorders in the early 1970s and created what might have been the first television time-shifting system. When he received the weekly *TV Guide*, Hefner would mark all the shows that he might want to see, and a full-time employee would individually tape each program on the reel-to-reel Ampex machines. Whenever he wanted to watch a program, any time of day or night, Hefner would call the video control room from one of the bedrooms and request that a western, or a classic film noir, or whatever else he might be in the mood for be routed to his television set.[6] Though the cost of Hef's Playboy mansion setup was out of the reach of most Americans, a handful of other enthusiasts set up professional machines in their homes, most notably the Sony U-Matic, which many television stations had begun using to edit newscasts and other programming.[7] Of these early home video hobbyists, many had professional connections to the media, whether working in a television station or, in one case, teaching English and mass media at the high school level.[8] Others, like many computer enthusiasts of the time, had access to video equipment through work or school.[9]

When Sony released the Betamax in North America during the 1975 holiday season, some hobbyists were waiting with bated breath—Marc Wielage, for example, was working at a TV station in Florida and had been dreaming of owning a home video machine since high school, when he read an article about Hugh Hefner's video equipment: "I contacted one of the big New York firms about buying an open reel VCR and using that to record programs off the air, because I worked nights. I wanted to be able to have a VCR that would turn itself on, record a show off the air and then shut itself off automatically." This wasn't possible with existing three-quarter-inch equipment, so when Wielage first heard about the Betamax he thought, "That's it. Eureka . . . Now I can have Hugh Hefner's setup at home!"[10] Other hobbyists were already involved in film collecting, and saw the Betamax as a more economical way to continue their hobby, while still more early Betamax owners had no previous experience with the industry or film collecting, and simply thought that the machine's time-shifting ability would be useful (especially those who worked nights or odd hours).[11]

But once people had the machines in their homes, what would they watch? There was essentially no mainstream prerecorded tape market in these frontier Betamax days. The early advertising campaign for the machine made no mention of renting or buying prerecorded tapes or, for that matter, watching home movies shot with a video camera—instead, the ads emphasized its ability to record television for later viewing, which Sony dubbed "time shifting." According to Sony, Betamax owners could make sure that they never missed a chapter of a TV miniseries due to a business dinner.[12] They could also watch two football games that were broadcast at the same time,[13] get more sleep,[14] and even add hours to their day, since "The hours you spend watching television are often the hours you could be using to get other things done."[15] One thing the ads never mentioned, however, was buying or renting prerecorded tapes. Because the major studios were not in on the development of the technology,[16] they were caught flat-footed by the introduction of this new machine, not having even figured out whether they wanted to release their movies and television programs on tape, much less how to do so. Thus, for the first two years that the Betamax was available, the only prerecorded tapes which users could legally buy were either public domain or pornographic.[17] The result of this emphasis on time-shifting was that if a video hobbyist wanted to watch, say, *The Prisoner*, he had to tape it off of the air himself, or get a copy from someone else who had taped a copy.[18]

Lugging the Betamax

The easiest way for a video enthusiast to get a copy of a program was to record it off the air, and for many hobbyists the most anticipated piece of mail was the weekly *TV Guide*. However, recording off the air meant that one had to wait until a given show or movie was broadcast, which could be frustrating for a budding enthusiast who really, really wanted to see, say, "that episode of *Mary Hartman, Mary Hartman* where she shows up at Sgt. Foley's house."[19] The solution for hobbyists was to extend the reach of their recording devices beyond their local broadcast stations.

For the extremely wealthy, one option was an Earth Station (later known as a satellite dish), which could pick up broadcasts of the new cable TV channels then in their infancy. Home Box Office (HBO) began broadcasting its signal by satellite in 1975, and was free for the taking so long as you could afford to spend upwards of $10,000 for a satellite dish (HBO didn't start scrambling its satellite signal until 1986).[20] Needless to say, this was out of the price range of most enthusiasts, even those who could afford to pay the $1295 suggested retail price for the first stand-alone Betamax, and it would be years until cable was rolled out in many U.S. cities.

With HBO and other pay cable channels initially out of reach at home, some early Betamax owners were resourceful: one advised in 1976 that "those of you with access to such things might consider taking your Betamax to a hotel that has closed circuit movies and helping yourselves to an evening's worth of film fare. For those so inclined, this would also be a possible way to secure 'XXX' type material, which is being run in motels in certain parts of the land."[21] Years later, the use of VCRs to record hotel movie feeds became a trope used by the Motion Picture Association of America's Piracy Office to characterize the most egregious film pirates.

If a particular program weren't available on a local channel, VCR owners would occasionally go to the extreme measure of bringing their Betamax somewhere that it would be broadcast: "I came out [to California in 1978] for a Grateful Dead concert at the Winterland Ballroom and they were broadcasting the Dead on TV, the whole show. So I brought my machine out . . . it had to be seventy, eighty pounds at least. I was going to the show, but a friend of mine wasn't going . . . and I trained this guy on how to [record the show], showed him how to load the tapes . . . So anyway it came off without a hitch, you know, that was the only time I ever took my machine out of state to do any recording, but that tape also wound up being a goldmine too, a nice recording and video of a Dead show. That was one of the first things I started circulating around."[22]

Other VCR owners were lucky enough to have connections that brought them copies of mainstream movies, whether through the entertainment industry[23] or from bootleggers at the local pub.[24] Sometimes, such connections were made by chance: "About '77 or '78 . . . I met a guy from the Air Force that had access to what they call the AFEX Library or the Armed Forces Radio and Television and he gave me a collection . . . of all the Outer Limits episodes."[25] If a hobbyist was lucky, the electronics salesman who sold him or her the VCR might be able to help out, though not officially—as one salesperson recalled, "It was something that, on your day off, they could call you or come by where you live if you were willing to deal with them . . . it wasn't something that you could have or should have done at work."[26] In the end, though, such inside connections were few and far between, and most early VCR owners were stuck with what they could record off the airwaves.

Building Social Networks

Ultimately, there was only so much that an individual videophile could tape on his or her own, and video enthusiasts quickly realized that their results would be far better if they pooled their efforts. On the simplest level, individuals could bring their VCRs over to the home of another videophile and dub copies of tapes from one machine to another. Ray Glasser remembers, "[In 1976] I used to lug my fifty-pound SL7200 over to [a friend's] house, we'd go up in his attic, where we had some room, and we dubbed tapes, and I'd bring it back that night or the next day to my house."[27] Even the simple act of connecting two Betamaxes, however, required a little bit of tinkering. There were no instructions on dubbing available from Sony (many electronics salesmen even claimed that it was impossible to do so), and the earliest Betamaxes didn't have the video or audio inputs that make connecting modern recorders so easy. Hobbyists had to improvise, connecting their machines via radio frequency (RF) antenna cables and tricking one VCR into recording the signal of the other as though it were being broadcast over a television channel[28] later models eventually included direct video inputs and outputs.

In 1976, a "36-year old staff director for one of the legislative staffs of the Florida House of Representatives" named Jim Lowe bought a Sony Betamax.[29] He had been involved with 16mm film collecting "on a very low level," and more generally was a collector of comic books, fantasy art, and in his words, "any number of obscure corners of popular culture." Seeing video as similar to his other collecting hobbies, Jim placed an ad in a 16mm collector publication called *Movie Collector's World*, which essentially read "I just bought a Sony Betamax recorder, it's the greatest thing ever, who out there might be interested in exchanging tapes?"[30]

A handful of people responded to that ad, and later that year Jim sent seven of them the first issue of *The Videophile's Newsletter*. In that first four-page issue he wrote about collecting videotapes in short, hand-typed paragraphs: "What I would most like to do is trade tapes with those of you who are willing to keep an eye out for my wants, while I will, of course, do the same for you. At present I have neither the time nor the inclination to be a taping service (e.g. tape every episode of certain shows every day for someone). As much as possible I would like to keep things on a strict hobby-type nonprofit basis. I have no desire to gouge you for the opportunity to see shows that are of interest to you if it is within my ability to bring them to you."[31] In the following two pages, Jim outlined the specific shows he was interested in, even going so far as to include the *TV Guide* listings from their original air date.

Over the next few months, the subscription list for *The Videophile's Newsletter* grew to dozens, then hundreds. Subscribers would send in their wish lists and contact information, first as a few lines of text to be printed and later as quarter-page ads. Ray Glasser, who had already been dubbing and trading tapes with a nearby friend, later wrote about the experience of discovering *The Videophile's Newsletter*:

I was thrilled that there were people who shared my interest; I subscribed immediately, and ordered all back issues. Along came March 1977, and I decided to run a full-page ad with my friend . . . decided to go the whole route. After racking my brain for a few months, I thought of all the TV shows and movies that I hadn't seen in years (but remembered), and put them down in our double Ad. And waited.

Along came April 1, 1977. Received the Newsletter with our Ad, knowing that we would reach at least one hundred other Betamax owners. Elated just to see it, I anxiously hoped to get a few letters. Beginning a week later, my mailbox began to fill up with something other than bills. It was beautiful! Then came the mindblower: that Saturday night, my pal and I were sitting over my apartment watching Forbidden Planet (which I finally got a hold of via another TVN subscriber . . .). The phone rings. A faint voice from the other end claims he's calling from San Francisco, California (!!!), saw my ad in the Videophile's Newsletter, and wants to trade. Half an hour later, (as if this weren't enough to send me skyrocketing), the phone rings again . . . This one's from Miami, Florida. And so on. Total that night, four calls. And once in a while, the phone still rings. Unbelievable!!! . . .[32]

Glasser's story was typical, and by the end of 1977 hundreds of videophiles were trading, taping, and corresponding with each other: "We all went broke on long distance phone calls and buying blank tapes."[33] In some cases, hobbyists sent "video letters," personalized videotapes recorded for one another using video cameras in which they would talk about their lives and give virtual tours of their homes.[34] Taping parties were common: "a dozen people or more would bring over their VCRs, they would daisy chain them together with one master machine feeding a half dozen or a dozen slave machines . . . If you got a copy of a brand new movie, oh you were like the hero of the month, and everybody would go over, crack open a beer, and do that."[35] For gatherings on an even larger scale, *The Videophile's Newsletter* later published an article about how to organize a video convention like the one described at the beginning of this chapter, suggesting that organizers post a "dubbing schedule" of when particular movies or shows would be available for copying and reminding organizers that "these VCR's and TV's suck a **lot** of juice, and you must allow for that, paying attention to the number of wall sockets, fuses and heavy duty extension cords are a must. Overloaded circuits go 'pop' and multiple machines suddenly go dead. It's an awful feeling."[36]

As a social practice, the general protocol for trading was straightforward whether in person or by mail, bearing a close resemblance to the preexisting norms for trading concert bootlegs of the Grateful Dead and other bands. Jim Lowe advocated a "one for one" system, with videophiles making even trades of one show for another, but if a trader didn't have anything to offer in return, he might build up his collection by offering "two for one," two blank tapes in return for one containing the desired program. The whole system was held together by good will, and particularly egregious bad-faith traders were outed in the pages of *The Videophile's Newsletter*.

Figure 1.2
A table of daisy-chained VCRs at the 1979 Video Collectors of Ohio Convention. (Video stills courtesy of Ray Glasser.)

As mentioned earlier, *TV Guide* was an essential tool for videophiles, letting them know when to program their VCRs to capture specific episodes of shows and to budget their tape accordingly. From the first issue of *The Videophile's Newsletter*, the importance of *TV Guide* was clear, and one of the most central concerns of its first few issues was how to trade *TV Guides* from different parts of the country. In November 1976, Lowe advised, "When sending a *TV Guide* to someone, you can save a little on the postage by taking out the center pages and discarding the rest."[37] In the next issue, he revealed that it was possible for videophiles to subscribe to editions of "the magazine we know and love so well" from any of the ninety six viewing regions of the U.S., and that he had in fact already subscribed to those for central Florida and metropolitan Los Angeles. This was such a big announcement that Jim actually included a copy of the response he'd received from *TV Guide* in that issue of *The Videophile's Newsletter*.[38]

In addition to sending out copies of their local *TV Guide*, videophiles also maintained lists of the tapes in their personal libraries. In its first year, *The Videophile's Newsletter* published subscribers' library lists, and as their collections grew some hobbyists published their own, updated every month or two, sent out as an informal newsletter to trading partners.[39] Augmenting the larger network of *The Videophile's Newsletter* subscribers, such smaller networks flourished, linking trading partners by phone and mail.

"You don't want to see Morris the Cat in the middle of Casablanca"

Whether recording for oneself or for a fellow enthusiast, one of the biggest difficulties of early off-the-air home video recording was getting a "clean" copy of the program. For many of the early videophiles, commercials were an unwanted intrusion that detracted from their enjoyment of the program. Moreover, with the cost of blank tape so high (in 1976, a one-hour Betamax tape could cost anywhere from $15 to $30), the fifteen minutes of commercials in an hour of network broadcasting were too expensive to waste; quite literally, time was money.

In order to create a copy of a broadcast program without commercials, videophiles would go to extreme lengths. Since most VCR owners initially only had one deck, the only way to get a clean copy of the program was to edit out the commercials in real time, starting and stopping the VCR as the program was being broadcast. This activity flew in the face of the advertised nature of the Betamax as a time-shifting machine—instead of using their VCR timers to record shows they were unable to watch, videophiles would sit in front of the television, watching the program intently while recording it.

Such real-time editing influenced the very placement of the VCR in the viewing room. Because the VCR was tethered to the television set by a short coaxial cable and VCR remote controls were still years away, editing the commercials out of a program often meant sitting within arm's reach of the television, not the most comfortable way to watch a program. One of the early concerns of videophiles, in fact, was how far the VCR could be from the television set, and thus how long the connecting cable could be without suffering signal degradation, explicitly so that "you [could] put the unit next to your chair and facilitate editing out commercials."[40] Many videophiles configured their homes in this way, with the VCR controls easily accessible from an armchair across the room from the television set, and some kept this spatial layout even after the introduction of remote controls.[41]

The actual process of editing out commercials was an oft-discussed topic, and tips for doing so were in high demand. Though simply pressing the pause button on the recorder when a station went to commercial was the easiest option, the result was a moment of commercial before the edit, and in the words of Jim Lowe, "you would rather miss a second of the show than to have a second of Morris the cat in the middle of Casablanca."[42] Thus, users developed increasingly elaborate processes to ensure the most "professional" edits possible, passing these editing skills back and forth through vividly written explanations.[43] Demand for cleanly edited programs was so great that by mid-1979, at least three competing companies were marketing devices that would automatically pause a recording VCR when a program cut to commercial.[44]

This emphasis on "professional" or "clean" edits implies a certain kind of expertise as well as a visible concern with the product of videophiles' labors. Though they might not have used the word themselves, their attention to detail and the ways in which videophiles wrote about their taping conveyed a sense of authorship. While they simply saw themselves as trimming away the unnecessary fat, these videophiles were in fact reshaping the broadcast feed to match their idealized vision of what the program *should* be, much as a turntablist or recording engineer might strip out certain sounds in order to bring out others.

At times, this bricolage was even more explicit. One videophile wrote in 1976, "I stayed up until 5 am along with Barbara Walters and Harry Reasoner to record the highlights of election night . . . I condensed the whole business down to an hour and still was able to catch early returns, both speeches and their respective half-hour 'last pitch' shows from the night before." He later wrote, "When I tape condensed versions, I leave on opening and closing credits and tape the piece I want, minus commercials, in the middle. It makes for a nice half-hour show."[45] Another early VCR user recalls editing together music videos in real time after MTV began broadcasting in 1981: "I would literally spend hours recording my favorite videos off of MTV . . . I'd just sit there for four or five hours at a stretch. Martha Quinn would tell you what was coming up that hour so you knew whether you really needed to sit there, but if you wanted to record [Michael Jackson's] *Thriller* you had to wait for it."[46]

Sometimes the boundary drawn by early videophiles between the desired program and unwanted material like commercials was much more flexible than one might expect, and one videophile's garbage might be another's treasure. One hobbyist pleaded the case for commercials themselves in the name of posterity: ". . . perhaps the preservation of selected commercials might be more rewarding [than archiving television shows themselves] . . . A reel or two of these might be of greater value in 2076 than ever so many episodes of *The Mothers In Law*."[47] The writer then redrew the boundary between useful and worthless programming, arguing that some commercials are "best buried and forgotten," but that others are interesting and important, and as worth keeping as the program they accompany. The process for recording these "worthwhile" commercials was the same as that developed for editing out commercials, just inverted.

In a particularly ironic twist, Jim Lowe proposed that the best way to preserve commercials without wasting precious tape was to record them onto "the 10–12 minutes that are left at the end of a K-60 tape after an hour (minus commercials) of other material that you are intending to keep is already on the tape."[48] Thus, a video enthusiast might edit the commercials out of a program, only to later record the commercials from a different broadcast, producing a tape with both program and commercials in the same proportion as the original broadcast, but now segregated from one another.

At other times, there was essentially no distinction made between commercials and program at all. Looking back, Ray Glasser said, "I can't stand commercials. Unless I'm archiving them, and say, 'an originally broadcast show with original commercials.' To some guys, this was gold."[49] In this case, the boundary between the program and the commercials is erased entirely. Some collectors specifically sought out recordings of programs as they were originally broadcast, leveling the distinction between "program" and "commercial" in favor of a more symmetrical view wherein both are essential to produce the ideal version of the show as it was originally seen, unmediated by the VCR. In the end, for those who thought commercials worthless and a hassle to edit out, the expectation was that the long-promised videodisc formats would satisfy "the desire for most movies . . . and that TV tape will be useful primarily for preserving TV shows themselves as well as obscure films . . . that no respectable video disk maker is ever likely to produce."[50] Until then, many videophiles' VCRs would remain firmly planted at the side of their armchair.

Opening up the VCR

As for the VCRs themselves, users were tinkering with them from the day they first showed up in stores. The first Betamax sold by Sony in the fall of 1975 was the LV-1901, which combined a television set and a VCR in one massive cabinet. One hobbyist remembered, "When I originally saw the console, I examined it very carefully and I realized that all they've done is taken a VCR and just packed it in a big wooden box and tied it to a TV set."[51] Though I haven't come across any cases of users buying the LV-1901 and actually removing the Betamax, hobbyists had mentally done so before Sony even announced the introduction of a stand-alone deck.

For hobbyists, *The Videophile's Newsletter* was a reliable way to connect with other VCR owners and trade information and technical tips as well as tapes. Within the first few issues, two readers in particular, Joe Mazzini and Marc Wielage, established their expertise with the technical details of video by answering every technical question Jim Lowe posed. Mazzini's "U-Matic Notes" (which later became "U-Matic and Beta Notes" after he purchased a consumer-level machine) was a fixture from the second issue, comprising the last few pages of each magazine. Wielage, who had professional experience in film and television, was soon named technical editor, and frequently contributed articles on the newest technologies.

On one hand, many of the early videophiles seemed to possess an enthusiasm for the technology itself, and no issue of *The Videophile's Newsletter* seemed complete without pages of reviews comparing the latest models of VCR and VCR accessories, usually accompanied by a two-inch photograph of the machine in question. Often manufacturers' product brochures were reproduced in their entirety, and every six months brought a feature story on the newest technology that debuted at the most recent Consumer Electronics Show.

Some videophiles, however, were interested in much more than just reading about the technology—they wanted to see how much they could do with it. In his first few issues, Jim Lowe posed questions about wiring and basic functionality, which Wielage and Mazzini quickly answered, and the discussions of what one could do with the Sony Betamax and other recorders were mostly limited to using its keys and buttons in different combinations. Tips ranged from how to best edit out commercials to a way to squeeze extra seconds out of videotapes by rewinding when first putting the tapes into the VCR before recording.

Within a year, videophiles were being encouraged to literally open up their machines and tinker with the circuits. One of the first examples of this sort of technical advice related to the "muting circuit" on the Sony Betamax. When Sony created the Betamax for the American consumer, particular care was paid to ensure that the image it displayed on the television screen was as clear as it could possibly be. One of the technical ways to ensure this was to put a muting circuit on the device that would blank the screen when the image was unstable (for example, during the first few seconds of playback, or while pausing). For many videophiles, however, this muting circuit amounted to an infringement on their control of the technology, and several ways of disabling it soon appeared in the pages of *The Videophile's Newsletter*.[52] As new VCR models were released by Sony, RCA, and other companies, invariably one of the first things to appear in print would be an explanation of how to defeat its muting circuitry.

Muting hardware aside, *The Videophile's Newsletter* carried explanations of everything from wiring together multiple VCRs for dubbing to enabling the auxiliary inputs that Sony had buried deep within the guts of the Betamax.[53] One enterprising user began splicing the magnetic tape inside cassettes as if it were film, while another rebuilt an old reel-to-reel video recorder to play Betamax tape.[54] The appreciation for technical expertise was so great among some videophiles that they celebrated Rick Redoutey, a technician at the Sony service center in suburban Detroit who won a national competition, and went on to be named the "number one Betamax technician in the world."[55]

Tips weren't confined to VCR hardware, either—one reader wrote in explaining how his friend had modified a new video disc player to play with the top off, with the added bonus that "the laser is visible through the *other side* of the disc!"[56] Readers were even schooled in the nuances of television broadcasting, in connection with common problems experienced by early enthusiasts:

... the cause of the color "Snow" on your set during a black and white recording is due to the station showing a black and white movie on a color film chain, with the "color burst" of the camera kept on ... Call the offending station and ask for Master Control, and mention the problem I explained; I've done it before locally, with the result usually being the poor engineer apologizing profusely and a proper picture popping on the screen seconds after my call.[57]

This particular bit of information reflects an understanding by video enthusiasts that their VCRs worked as part of a larger sociotechnical system, and that in addition to tweaking the settings on their decks they occasionally needed to tweak other people within the network.

The high levels of technical knowledge concentrated in videophile networks resulted in a remarkably high level of interaction between users and the manufacturers of their equipment. In the first years of the video industry, salespeople in electronics stores sometimes knew far less about the goods they were selling than did their videophile customers,[58] and many enthusiasts went straight to the source, ordering product brochures, repair manuals, and even production guides directly from the hardware manufacturers. When doing so, videophiles sometimes assumed the role of experts; for example, Joe Mazzini advised videophiles interested in a particular guide to program production for the as-yet unseen Laserdisc system to write to MCA and "mention that you're interested especially in the programming abilities of the disc"—in other words, imply that you're a manufacturer, rather than a hobbyist.[59]

Tinkering and Hacking

This sort of tinkering with a new technology was of course far from novel. In the early days of wireless telegraphy, skilled amateurs (traditionally cast as "boy operators") ranged through the ether, dominating the development of the new medium to the point of direct conflict with the U.S. Navy.[60] Later, once the medium had shifted to voice and music broadcasts, hobbyists known as "*dx*-ers" (and later, "hams") tweaked their radio sets to receive signals from around the country and around the world.[61] Audiophiles have pushed the boundaries of both their hi-fi stereos and their own ears in their

Figure 1.3
Videophiles at the 1981 Consumer Electronics Show in Las Vegas: Ray Glasser at the *Videophile Magazine* table (top), and Art Vuolo interviewing Dick Unger of Sony Electronics (bottom). (Video stills courtesy of Ray Glasser.)

quest for the purest sound.[62] The early days of personal computing were dominated by enthusiastic hobbyists, concentrated in groups like the famed Homebrew Computer Club where members built both personal computers and an industry from the motherboards on up.[63] Outside the realm of information technology, early rural automobile owners opened their own black boxes, reconfiguring their Model Ts to serve as everything from plows to washing machines, and decades later hot rod culture reimagined the car not just as a means of transportation, but also as an aesthetic canvas on which to paint literally (with flames and pinstripes) and figuratively (with chrome and steel).[64]

If technological literacy can be defined as the knowledge of how to use a given technology, then it might be said that such tinkerers possess a certain technological fluency—the ability not only to *read* meanings of the technology, but also to *speak* new ones. Such users don't simply draw on black-boxed knowledge; they unpack it to suit their whims. In the abstract, these tinkerers occupy a cultural space in between producer and consumer; Kristen Haring points to the ways in which these hobbyists invoke the language of "amateurs," often with problematic results.[65]

Such tinkering has most often been described as a solitary activity, and studies of tinkerers like hams and *dx*-ers often center on the relationship between an individual's tinkering and his identity construction, particularly as it relates to the renegotiation of masculinity in the twentieth century. While valuable, in the case of hobbyist *communities* this focus on identity construction can describe the trees, but not the forest. An enthusiast community overflows with newsletters, leaflets, meetings, mailings, and other formal and informal channels that mediate information between members.[66] And at the heart of all this communication is often the cultural norm that it is *good* to tinker, and moreover that it is *even better* to share your tinkering with other hobbyists.

One canonical example of such a hobbyist community can be found in computing's early days. In their actively *social* engagement with technology, many computer enthusiasts adhered to what Steven Levy calls the "Hacker Ethic." The culture of early computer enthusiasts and hobbyists, Levy writes, was based around "a philosophy of sharing, openness, decentralization, and getting your hands on machines at any cost—to improve the machines, and to improve the world."[67] To Levy, this ethic was neatly encapsulated in the verb "to hack," the act of finding an elegant solution to a problem using the technological tools at hand. Sherry Turkle comes to a similar conclusion, explaining that "a central organizing theme in hacker culture . . . is 'The Hack.' It is the holy grail."[68] For both Levy and Turkle, "hacking" goes beyond mere tinkering in its aesthetic appreciation of the elegance and beauty of a particular bit of tinkering, as well as its normative assertion that to hack is fundamentally good.

Moreover, such hacking extended beyond mere tinkering with the computer's technical details to altering its social context. As one of Levy's hackers explained, the MIT students who first began programming on the PDP-6 would go "lock-hacking":

. . . there were administrators who would have high-security locks and have vaults where they would store the keys, and have sign-out cards to issue keys. And, they felt secure, like they were locking everything up and controlling things and preventing information from flowing the wrong way and things from being stolen. Then there was another side of the world where people felt everything should be available to everybody, and these hackers had pounds and pounds and pounds of keys that would get them into every conceivable place.[69]

In this case, it wasn't just a program or a lock that was being hacked—these MIT students were actively tinkering with the system of social norms and relationships surrounding the computer in order to reshape the Hughesian sociotechnical system in which it existed.[70] Much like the technical hacking by which MIT hackers would constantly revise and improve code (regardless of whether this was against the rules), this sort of social hacking wasn't done with any malevolent intent, but simply to make the technological system as a whole run more as its hackers thought it should.

Though this aesthetic component of hacking is central to its definition, there is another aspect of hacking that is just as key, though more often overlooked. Hacking as an act does not exist without a larger community of hackers, and the aesthetic judgment of a "good hack" is made not just by the individual, but by the larger group of her peers through in-person demonstration and distant replication. As Turkle writes, "The ideal of the hack suffuses the hacker culture. It embodies shared values and passions. And, of course, it is the centerpiece of hacker rituals." Hacking, as opposed to tinkering, necessarily points beyond individual actions toward broader social relationships.

Turkle explains that The Hack "is a concept that exists independently of the computer,"[71] and if we take "hacking" to mean "tinkering based on *community* norms of elegance, openness, information-sharing and decentralization," it is readily apparent that hacking is a vital element of many enthusiast communities. The term can be used anachronistically to describe the activities of radio dx-ers who published articles on how to push reception boundaries even farther than the Navy thought possible, as well as the telegraph operators who, ignoring the transcription mechanism Morse designed into his apparatus, discarded compositors and began to directly tap messages to each other using the telegraph key.[72] In addition, an analysis of a hacker culture doesn't need to limit itself only to how its members tinker with the technical details of the object of their enthusiasm, but also can examine the ways in which they hack the social milieu in which it exists.

While they wouldn't have used the word themselves, members of the early videophile community quite consciously hacked their VCRs, as well as the VCR's broader sociotechnical context. The active and vigorous debates that took place in the pages of *The Videophile's Newsletter* over the best way to disable the Betamax muting circuit recall meetings of the Homebrew Computing Club, and phone calls to television stations imply a level of sophistication about the workings of the larger sociotechnical network comparable to phone phreaks' use of blue boxes.[73] In a sense, we can understand the network of phone calls, classified ads, and *TV Guide* subscriptions as a form of hacking, produced by tinkering with traditional social relationships, with an underlying normative belief that greater access to television broadcasts was inherently *good*.

In foregrounding the normative aspects of some tinkering, the use of hacking as a larger analytic category points toward the ways that tinkering may or may not fit within the broader cultural context. Helen Nissenbaum, for example, argues that computer hackers were viewed with "grudging admiration" through the 1960s and 1970s. Computer hacking acquired its contemporary connotation of illegal deviance not because hackers themselves changed, but because their context did. In the light of a growing corporate reliance on information technology and the attendant politics of information as property, Nissenbaum explains, computer hackers were reframed from heroes to hooligans.[74]

We can understand the famous *Universal v. Sony* court case as a similar attempt to frame videophile hacking as deviant and a violation of traditional standards of property and decency.[75] Universal Studios claimed that Sony's VCR allowed users to infringe on its copyrighted television broadcasts, and filed suit in federal court to stop Betamax sales. In the winter of 1977, Jim Lowe was subpoenaed by Universal Studios and Disney, and in February

1978 was deposed for more than four hours about his experiences as a video hobbyist and as the publisher of *The Videophile's Newsletter*. He immediately sent out a letter to everyone whose name and address had appeared in the *Newsletter*'s pages, advising them that though he had refused to turn over his subscriber list, the Universal/Disney legal staff might contact them as well. Lowe also proceeded to print selections from the deposition transcript in the next issue of *The Videophile's Newsletter*, including a copy of his subpoena on the front cover.[76] A year later, Marc Wielage was also subpoenaed and was called to appear at the trial itself, where he testified about "what kinds of programs I taped off the air, why I eliminated the commercials, how I did so, how often I watched tapes, how large my library was, and how *The Videophile* had been started and what its purpose was."[77]

Throughout the trial, Jim Lowe's magazine was held up as the tangible embodiment of a network of videophiles, and Lowe and Wielage as representative of Betamax owners everywhere.[78] The very norms of openness and information sharing that had driven them and others to create a nationwide videocassette trading network were described by Universal's attorneys as particularly egregious examples of disrespect for the studios' intellectual property. The case worked its way through the appellate courts through the late 1970s and early 1980s, and by the time the U.S. Supreme Court found in favor of Sony in 1984, the normative point was moot—regardless of its legal status, using the VCR to time shift (and even archive) television broadcasts had been decided to be morally in the clear. Unfortunately, though their hobby had finally been validated as legal, by that time the videophile community had faded to but a whisper.

The Decline of Videophilia

One can chart the rise and ultimate fall of videophile culture through the pages of *The Videophile's Newsletter*. Initially sent out to seven people in the fall of 1976, the magazine's circulation rose to thousands by late 1978, thanks in part to mentions in newspapers and magazines like *Playboy*, *The Wall Street Journal*, and *Money*. Jim Lowe, still working a full-time job in the Florida Legislature and publishing *The Videophile's Newsletter* in his spare time, decided to "take the plunge and turn what has been a nonprofit publishing hobby into a serious . . . publishing enterprise."[79] Asking his readers to invest in return for extended (and even lifetime) subscriptions, Jim raised enough money to open an office with several employees, and with the January 1979 issue a sheaf of stapled-together pages called *The Videophile's Newsletter* became *The Videophile*, a glossy magazine with a full-color, professionally photographed cover.

At its peak, *The Videophile* had around 8,000 (mostly Caucasian, mostly male) subscribers and was sold at newsstands around the country.[80] Its writers were seen as experts on the burgeoning field of home video, and appeared everywhere from industry panels to radio and television talk shows.[81] *The Videophile* itself established more of a presence at larger events, with a table at the 1978 National Film Convention, then a booth at the 1979 Consumer Electronics Show.[82] The hand-typed, friendly ads from enthusiasts and local dealers that used to fill the magazine's pages were gradually overshadowed by slick, polished ads from major manufacturers and retailers (though the mini-ads posted by collectors looking to trade remained). In 1982, however, things took a turn for the worse for *The Videophile*, as a trademark dispute over the use of the word "videophile," bad luck with distributors, and rising printing costs sapped the magazine's finances, while magazines like *Video* and *Videoplay* from major publishers cannibalized Jim Lowe's efforts to grow his subscriber base. Within a year, the magazine became unsustainable and folded.

A year later, Marc Wielage and another videophile named Rod Woodcock got hold of *The Videophile*'s subscriber list and launched a new magazine, *Videofax*, intending to cater to "the hardcore *videophile* audience" (emphasis theirs). In their introduction to the new magazine, while praising *The Videophile* and emphasizing that they were going to return to "the spirit of the old magazine," Wielage and Woodcock emphasized their interest in the technology of video:

We'll be concentrating strictly on hardware topics, along with nuts and bolts articles considered too technical, too esoteric or subjective for the other mainstream publications; we won't review movies, tell you how to tape your kids' birthday parties, or how to hook up your VCR to cable TV. We're going to assume you already know enough about all those basics. For those of you who don't have experience with home VCR's, we'll tell you right up front that you may be happier reading the newsstand magazines than us.[83]

Unlike Lowe, who tried to reach out to newcomers to video until the last issue of *The Videophile*, Wielage and Woodcock explicitly targeted highly technical enthusiasts as their audience, in the process hardening the very

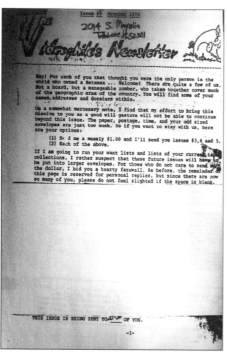

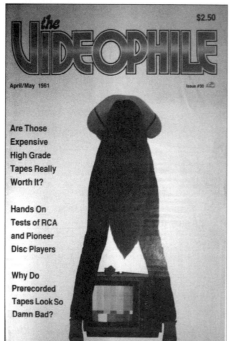

definition of "videophile." As home video became an increasingly common consumer phenomenon, it became clear that most users did want to just watch movies and videotape their kids' birthday parties, and weren't interested in reading anything more than their instruction manuals (if even those). *Videofax's* increasing emphasis on technical details over the course of the mid-1980s reflected the increasing ghettoization of video hobbyists from mainstream VCR owners.

This seems to be the fate of most enthusiast cultures—they carry within themselves the seeds of their own irrelevance. The more successful enthusiasts are at pushing their hardware beyond its manufacturers' intentions and evangelizing its possible benefits to the general public, the more likely it is that their hobby will become a major industry, bringing with it their own marginalization. Consider the history of radio: as radio receivers became commonplace, the popular adulation of the boy operator was replaced by a slight wariness of the oddball ham.

Earl Muntz, at one time the largest home video dealer in southern California, once told Marc Wielage, "The biggest thing that's going to change the industry is when the VCR becomes as commonplace as a toaster." Later, Wielage understood what he meant: "We were writing about video as an esoteric, exotic, high-end medium for hobbyists who are into the technology. The problem is, once it got to the mass market level of the toaster, suddenly all the things that made it special were gone."[84] Once the wave had subsided, the enthusiasts who had ridden the crest of the new technology found themselves washed ashore on a crowded beach.

A particular aspect of the enthusiast character thrives on novelty and the excitement of the frontier, where the technology in question is full of potential and the knowledge of its secrets marks one as special. Once the VCR was a commonplace item, for many enthusiasts "it became too mainstream . . . the kick was gone, the original thrill was gone."[85] Moreover, as the prerecorded tape industry took off and one no longer had to record a movie off broadcast television to possess a copy, the skills accumulated by videophiles

Figure 1.4
Covers of *The Videophile's Newsletter, The Videophile,* and *Videofax*. Note the increasingly polished and mainstream imagery as *The Videophile* reaches for a broader audience, as well as *Videofax's* return to a more asexual, fetishized orientation toward the technical. (Courtesy of Jim Lowe and Marc Wielage.)

began to seem unnecessary. In a sense, while integral to selling it to a wider audience, the black-boxing of the VCR as an easy-to-use consumer good (like a toaster) rendered video hobbyists obsolete. While a handful continued trading tapes and meeting up at the Consumer Electronics Show every year, the practices of the growing number of VCR owners veered away from the tape-trading classifieds at the back of magazines like *The Videophile* and toward the next frontier, prerecorded cassettes.

2 "Hollywood in a Box": Reconstructing Videotapes as Transparent Media

In the early 1980s, an apocryphal story made the rounds among video store-owners concerning a hapless customer who brought his VCR back to the store where he'd purchased it a year earlier, complaining that it had stopped working. The storeowner looked it over, wondering if there had been some mechanical failure, but found none. Upon ejecting the videocassette currently in the deck, the storeowner found that it had been played and recorded over so many times that the magnetic tape had worn to the point of snapping. Handing the customer the tape, the storeowner asked if all of his tapes were this worn, to which the customer responded, "I didn't even know that piece came out!"

On its surface, this is a simple joke about technological ignorance, riffing on the idea that someone could be so clueless as not to realize that video tapes are interchangeable. However, the joke is only funny because it plays on the generally understood assumption that VCR use is intimately connected with the use of multiple tapes.[1] If the VCR were simply seen as a time-shifting machine (not unlike a contemporary TiVo), the joke wouldn't make sense—of *course* the customer never tried to change tapes, because to him the VCR was simply a black box that recorded and replayed television.

From its earliest days, the VCR was quickly understood as more than simply a time-shifting device. For the videophiles in the last chapter, it was a tool used to time shift and collect their favorite television programs, extending their television viewing into different geographic and temporal spaces, and blank videotapes were an integral part of the technology. However, as the technology became more widespread through the late 1970s and early 1980s, the VCR was increasingly identified not just with blank tapes, but also with a growing number of prerecorded tapes. For users who weren't hobbyists or enthusiasts, renting a movie on a weekend night would eventually become as commonplace an activity as listening to a record album or watching television; in the meantime, as the market for movies on videotape developed, the enthusiast emphasis on the cassette as a material technology faded. Videotapes themselves became understood less in terms of their technical nature as interchangeable parts of the VCR and increasingly in terms of the texts that they carried, thanks to the efforts of entrepreneurs who insinuated themselves between VCR manufacturers, Hollywood studios, and consumers.

Blank Tapes

Attempts to record video go as far back as British television pioneer John Logie Baird, who in 1927 recorded television signals as grooves on phonograph records. Not unlike his mechanical television, Baird's "Phonovision" never really took off, and for the first few decades of the television industry the standard way to record a broadcast, either for archival purposes or to rebroadcast later, was to essentially point a motion picture camera at a television receiver and capture the video image on film. Through the 1950s, the preferred system for doing so was the kinescope, and the best "hot kines" systems could shoot a broadcast in New York and have the film developed and ready for rebroadcast on the West Coast within three hours.[2]

By 1960, however, magnetic tape recorders were producing high enough resolutions that they were replacing kinescopes and other film-based technologies. Making its network debut by recording Nixon's 1959 Moscow debate with Khrushchev for later broadcast in the United States, the Ampex reel-to-reel videotape recorder offered instantaneous recording and playback in vibrant color. The Ampex machines, along with subsequent recorders by RCA and Sony, raced through tape at speeds of up to 15 inches per second, requiring massive reels just to record an hour-long program. The technical history of videotape in the 1960s essentially revolved around the question of how to slow down the recording speed (ultimatly shrinking the size of the tape reels).

In 1969, Sony developed the first color *cassette* video machine, using a cassette that measured 6 by 10 by 3¾ inches, and by 1971 they had revised their standard to ¾-inch-wide tape. The resulting videocassette recorder, which could play (and in later incarnations record) up to an hour of video with stereo audio, quickly became an industry standard, due in no small part to

Sony's intermanufacturer agreement licensing the U-Matic technology to Matshushita and JVC (Victor Company of Japan).

Building a market for the new technology, however, was more difficult than establishing the standard. Sony's initial approach was a sort of traveling road show, a well-choreographed demonstration of the videocassette player to audiences around the country ranging from the Institute of Electrical and Electronics Engineers and Hollywood boardrooms to major politicians and the military. The de facto English-language spokesman for Sony, a man named Peter Keane, had worked in the film industry since the early days of Technicolor and had been hired in large part because of his contacts in the film industry. Demonstration audiences tended to be very impressed with the new cassette technology, which appeared substantially more user-friendly than a hand-threaded projector or recorder: "16mm prints were not only bearing up in the machines, but scratching badly, and the videocassette was pristine, clear and bright with good color."[3]

Though later models included recording capability, the initial Sony U-Matic was a playback-only machine, and while audiences were impressed with the machine's capabilities, sales were lean. Keane was approached one day by Mike Tsurumi, a Sony engineer who was working on marketing the new machine: "[Tsurumi] said, 'What do we do? We can manufacture hardware, but we cannot seem to interest people in buying—there's no software.' So I came up with a scheme. I said, 'We should contact many different companies that do 16mm education and training programs for themselves, borrow a print, do a transfer, and go back to them with a "carousel" cassette and show them how simple it was to play back videotape rather than threading up a 16mm projector.'" Keane's hope was to develop "enough software to get the ball rolling," and by 1972 the educational film company Knowledge Industries had published a list of at least fifty or sixty programs available on 3/4-inch tape.[4]

The key selling point of the U-Matic was ease of use, and in packing the tape and reels inside a plastic case, manufacturers took a tentative step toward making videotape recording user-friendly enough to move from television studios and institutional settings into homes. The first magnetic video recorder explicitly intended for home use seems to have been the Ampex Signature V, which measured over nine feet long, included an Ampex reel-to-reel video recorder as well as a black-and white camera, television receiver, and home music center, and was featured in the 1963 Neiman-Marcus

Christmas Catalog for $30,000.[5] Needless to say, the Signature V (whose purchase price included a visit by an Ampex engineer to set it up) didn't achieve widespread market penetration. The move to cassettes was seen as essential to the creation of a consumer video recorder, and as described in the last chapter, some eager enthusiasts adapted the professional-grade U-Matic for use in their homes.

Interestingly, some of the U-Matic's earliest adopters outside of the television industry were operators of adult movie arcades, who had traditionally relied on never-ending loops of film to give their customers their twenty five cents' worth but found the video recorder more economical. Though the conventional wisdom that pornography fueled the early video industry is most definitely true, most people might be surprised to hear that the earliest adopters were adult bookstores and arcades rather than home users.[6]

In 1975, when Sony itself first offered a redesigned videocassette recorder for home use, the legacy of the U-Matic persisted. Like their ancestors, the first Betamax cassettes could only record an hour of video, an embodiment of Sony's assumption that the machine would primarily be used for time shifting broadcast television shows (which generally lasted an hour at most). Tape length was a crucial factor in the early skirmishes between Sony's Beta format and its rival, Matshushita's VHS—the story goes that RCA, the major producer of VHS recorders in the United States, refused to start manufacturing the machines until a tape could record not just an hour-long serial drama, but a full football game. Thanks to an innovative recording head design and thinner tape, the first RCA Selectavision VHS machine arrived in stores with a tape running time of two hours.[7] Though Sony had its own two-hour machine on the market soon thereafter, it never entirely managed to shake off the Betamax stigma of shorter recording time (albeit with higher image quality).[8]

As for the tapes themselves, the first cassette manufacturers for the Sony Betamax and RCA Selectavision were in fact Sony and RCA (respectively, of course). One consequence from the VCR manufacturers of the blank tape market's early dominance is that tapes were explicitly linked to the technological nature of the recorders. In 1976, a one-hour Betamax cassette was a piece of high-end technology that would cost you $15 if you bought in bulk,[9] while retail prices could go as high as $30.[10] Instructions both included with tapes and published in hobbyist magazines outlined the dos and don'ts of videocassette care, including how to store them (vertically, "like they were books"), how to protect them (ideally wrapped in a plastic sleeve or bag when not in use to protect from dust, ashes, and other contaminants), and how to clean them if need be.[11] Moreover, both tapes and recorders carried the same brand label, and manufacturers warned about the potential damage that inferior imitation tapes might cause to a machine's recording heads.

Initial studies claiming that VCR owners would want to purchase between one and five blank tapes proved to underestimate early adopters' desire to archive, rather than simply time shift, television, and chronic shortages plagued the industry's first two years.[12] The supply problem eventually worked itself out as companies like 3M and Fuji (and later TDK and Memorex), who were solely involved with magnetic tape fabrication, caught up to the VCR manufacturers.[13] When these competitors introduced their own cassettes to compete with Sony and RCA, they encouraged the vision of videocassettes as high technology. By emphasizing the technical nature of videotapes, separate from the recorders in which they were played, these newcomers made the case that their second generation videocassettes improved on the VCR manufacturers' earlier cassette technology. One Fuji ad in the summer of 1979 trumpeted: "Finer magnetic particles and more uniform distribution . . . Higher precision and stability in cassette housings, as well as the cassette itself . . . for greater viewing enjoyment and trouble-free recorder performance."[14] Not only would Fuji's improved videocassette technology lead to better images on the TV screen, the ad promised, it would actually improve the functioning of the VCR itself!

Recorders and Cassettes as Brown Goods

Videocassettes as a technology were also bound to recorders in the way they were sold to consumers. On both sides of the Atlantic, consumer electronics have historically fallen into two main categories, "brown goods" and "white goods."[15] In their study of the microwave oven and its sale in the British marketplace, Cockburn and Ormrod lay out the difference between brown and white goods—brown goods "are for leisure and entertainment," while white goods "are for domestic work and . . . for personal hygiene." Not only do white goods look different from brown goods, they are sold in different stores, or at least in sharply defined zones within the same store; as one of Cockburn and Ormrod's salespeople put it, "On entering the store the customer's left in no doubt which side is the fun side and which side is the boring, working appliance side." Moreover, the two categories are strongly

Figure 2.1
Stills from a 1976 Sony Betamax promotional video that explained how to use the VCR to time shift television. No mention was made of playing prerecorded videocassettes. The children in the last frame were brought onscreen to underscore the spokesman's assertion that family commitments would never make him miss a televised sporting event again—note the basketball game being recorded in the background. (Video stills courtesy of Ray Glasser.)

gendered, with the feminized white goods aimed towards housewives and other domestic workers in contrast to the highly masculine culture of brown goods.[16]

When Betamax and VHS recorders were introduced to the market, retailers and distributors immediately interpreted them as brown goods and placed them alongside home computers, video games, and other peripherals that reframed the television set as a general monitor rather than a broadcast

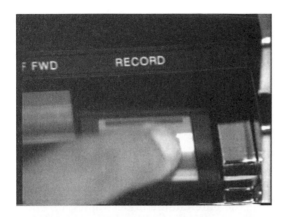

Figure 2.1
Continued

receiver.[17] When presenting the new technology of video recording to customers, many retailers followed the lead of the manufacturers, who across the board believed that as an integral part of the VCR experience, blank tapes should be sold right next to the recorders.[18] To that end, manufacturers required retailers to carry cassettes alongside VCRs—initially, it was often not possible to sell blank videocassettes without also selling recorders, simply because manufacturers like Sony wouldn't let storeowners buy one without the other.

Video recorders and blank cassettes were picked up not just by brown goods retailers, but also by the established distributors and manufacturer representatives who were already carrying television sets and hi-fi stereos.[19] These were middlemen in the truest sense, usually carrying product lines from many different manufacturers and consolidating them into a single, one-stop-shopping opportunity for storeowners. Under this system, a retailer might only have to deal with a handful of mediators rather than dozens of manufacturers, while at the same time, a manufacturer would only have to deal with a given mediator rather than hundreds of retailers.

Since VCR manufacturers produced the first videocassettes, these mediators handled both recorders and cassettes. Most of these distributors and reps had been selling hi-fi audio equipment and television sets from Sony, RCA, and other manufacturers for years, so it seemed natural to add a new television peripheral to their product lines. To them, both recorders and blank cassettes fell in the general category of home entertainment electronics, in which they had built up extensive expertise. This stable understanding of the relationship between videocassettes and recorders would soon be renegotiated, however, as the prerecorded videotape industry began to expand.

Prerecorded Tapes

The first attempt to sell videocassettes already recorded with movies and other programming predated Sony's Betamax by more than three years. In 1972, Cartridge Television, Inc. announced its Cartrivision videocassette recorder, which played half-inch magnetic tape off of reels stacked one on top of another in square cartridges. Due to a novel technical design, a Cartrivision cartridge could hold up to two hours of video, and the machines, produced by Avco, had the same basic functionality as the later Sony and Matshushita standards.

The real selling point of Cartrivision, however, was the integrated sale of both medium *and* message; the machine would be marketed hand in hand with prerecorded tapes of mainstream Hollywood movies and other content. The Cartrivision system included two kinds of tapes: black tapes that were for sale only and included instructional titles and produced-for-video features, as well as blank black tapes for television time shifting and use with the Cartrivision video camera; and red tapes, which were for rental only, including major Hollywood films like *Casablanca* and *Dr. Strangelove*. Hollywood studios (most notably Columbia Pictures, which became a partner in the venture thanks in part to the efforts of Peter Keane, who had left Sony to join Cartrivision), traditionally wary of anything less than complete control over the distribution and exhibition of their films, had been persuaded to license their titles to Cartrivision because the rental-only tapes could only be rewound using special in-store equipment (which included a tamper-proof counter for accounting purposes), with the net effect that a customer's rental fee covered exactly one viewing of a given movie, not a second more.[20]

Thanks to a partnership with Sears, Roebuck and Company, Cartrivision went on sale at eighteen Sears stores in the Chicago area in June 1972, and there was trouble from the beginning. The first recorders were built into a console with a television set (not unlike the first Sony Betamax), and salespeople at Sears seemed to be unsure how to present it to their customers: "They knew how to talk about furniture. They could tell you how to turn a TV set on and off. But they had never run one of these things. They hadn't been trained . . . it was a mess."[21]

Eventually Cartrivision introduced a stand-alone deck, but other problems kept cropping up. The prerecorded tapes were sometimes sold on different floors of the store from the recorders, and in some cases renters had to order tapes via UPS and wait days for them to arrive. The no-rewind functionality of the red cassettes proved an irritation to consumers. Moreover, the fabrication process for the cassettes (which involved sandwiching together a strip of blank tape and a master, then passing an electrical current between the two to create a duplicate of the original) resulted in such poor audio quality that once hand-assembled into the cassette, the audio needed to be erased and rerecorded.[22] Worst of all, the entire stock of prerecorded tapes began to spontaneously degrade in November for reasons never fully understood, and there was concern that the use of one of these corrupted tapes could damage the recording heads of a perfectly good Cartrivision recorder.[23]

After all this, the Cartrivision management paused, gathered their resources, and prepared for a do-or-die market test in California. Learning from their mistakes, the company established the rule that "no store could go on line unless [it] had trained the salespeople in demonstrating the hardware and unless a sufficient supply of cassettes was available and visible with the hardware."[24] Within a few weeks, the test began to show results, but by this point Cartridge Television was hemorrhaging money. Around this time, Avco executives on a trip to Japan saw a demonstration of Sony's next-generation Betamax video recorder and, though Sears was ready to start marketing Cartrivision on a national level, Avco decided to pull out as a manufacturer, leaving Cartridge Television bankrupt.[25]

Public Domain and Pornography

In the first half of the 1970s, there were periodic murmurings from the consumer electronics industry about new technologies that would bring prerecorded media into homes. By 1970, for example, RCA was already heavily invested in its family of SelectaVision video players/recorders, including magnetic tape players, the promising holographic tape player, and even an in-development videodisc player, and the company believed that programming would be vital to the success of any of these technologies. RCA had hired a television executive named Thomas McDermott to build a programming business from scratch, and he spent several years with a prospective budget of $50 million trying to do just that.[26] Thanks to other corporate losses, however, RCA retreated from the programming business, leaving McDermott's efforts out in the cold.

Another videocassette player wouldn't reach the consumer market until 1975, when the Sony Betamax would be advertised by its manufacturer only as a time-shifting device. Akio Morita, Sony's chairman, had explored potential deals with several Hollywood studios, and even announced a joint venture with Paramount Studios in 1976.[27] At the time, however, none of the Hollywood studios were willing to risk releasing their films on a format that didn't offer strict control over users, and their tentative partnerships with Sony fell through.

Soon after the Betamax was unveiled, independent suppliers began to offer their own tapes with prerecorded content for sale. One of the earliest was Sound Unlimited, a Chicago-area electronics distributor. Noel Gimbel, the owner of the company, had built his business from music retail to distribution through the 1960s and 1970s and was an enthusiastic early adopter who owned an Advent projection TV, a Sony Betamax, and the first Atari video game, *Pong*. As Gimbel tells it, one day at the office he was complaining that there wasn't much to do with video except record television, when Jeff Tuckman, a film buff who worked at Sound Unlimited, suggested, "Well, why don't we do public domain?"[28]

Tuckman had been an active collector of 16mm and 8mm movies for years. From the first film he obtained (a 16mm print of the Marx Brothers classic *Duck Soup*), he had been enchanted with the idea of owning movies that he could show to his friends. Film collecting was an expensive hobby, however, so when Tuckman (whose cousin had worked at Sears during the Cartrivision debacle) first heard about the Sony Betamax, he naturally saw the machine as a more reasonable way to bring movies into the home. Thanks to his collecting, Tuckman was plugged into the network of film collectors, and was well aware that the copyrights had expired on many early films and television shows, meaning that anybody who had a master print could make copies and sell them, royalty-free. A handful of enterprising suppliers with names like The Nostalgia Merchant, who had been offering these "public domain" films on 16mm and 8mm film, had begun to transfer them onto videocassette and sell them by mail order.[29]

At the same time, the pornography industry had been embracing the new medium of videotape. Pornographers had explored the potential of film from the earliest days of motion pictures, and the "stag film" had been a discreetly tolerated institution in American life since the 1920s.[30] In 1971, pornographer extraordinaire Reuben Sturman, who had spent decades building an adult bookstore empire, invented the "peep booth," a small room with a coin-operated movie projector loaded with short loops of hard-core films.[31] At the same time, adult movie theaters in urban areas (particularly the Times Square area of New York City) served as both sexual and social meeting places for gay men.[32] However, it wasn't until the theatrical success of *Deep Throat* and *The Devil in Miss Jones* in 1973 and the ensuing burst of "porno chic" that adult film broke from the underground into American popular culture.[33]

Thanks to the questionable legality of such films (not to mention the difficulties that distributing them across state lines might raise with federal law enforcement), adult filmmakers were far less invested in maintaining strict possession of their work than mainstream Hollywood. Sturman and other producers of peep-booth loops harkened back to the early Edisonian days of the film industry, when kinetoscope shorts were sold outright to the nickelodeon owners. As for the more narrative adult films, piracy was less a fear than a fact of life—*Deep Throat*, for example, was widely copied without permission. Rather than take action through the legal system, however, the film's financiers (who were connected with organized crime) simply had their "associates" visit theaters where bootleg copies of the movie were playing, offering theater owners the chance to continue showing the film in exchange for half of the box office receipts. As Eric Schlosser notes, "Few theater owners refused this offer."[34]

Thanks to their existence on the fringes of mainstream America, those in the adult film business seemed to feel freer to try new things, among them releasing their films on videotape. Mail-order houses began offering video-tape copies of recent theatrical sex films (as well as older films by directors like Russ Meyer and Radley Metzger) as early as 1976, advertising them in the back of film and video hobbyist magazines for up to several hundred dollars apiece. Just a few years later, in 1979, *Playboy* told its readers that "just about every top-quality X-rated movie made in the last several years can be legitimately purchased over the counter."[35]

Putting all of this together, Jeff Tuckman saw a business opportunity for a distributor like Sound Unlimited. Tuckman thought that he might be able to convince Sound Unlimited's record store customers to give videotapes a shot, and persuaded Noel Gimbel to give him $5,000 to start up a video distribution arm of the company. "I bought a handful of public domain and adult tapes," Tuckman recalls, among which were classics like *Night of the Living Dead* and *It's A Wonderful Life*, as well as some music-related titles that he thought might sell well to record stores.[36] The tapes themselves were "kept locked in a cabinet in the corner of our warehouse," a cabinet to which he had the only key and from which he would fill orders himself. Once he had an inventory, Tuckman began trying to sell his record store clients on this new idea of videocassettes, but they were not yet buying what he was selling. "They would come on the [Sound Unlimited] intercom every day and say 'We sold one videocassette today.'"[37] Less than three hundred miles away, however, another entrepreneur was hatching a plan that would soon have Tuckman's phone ringing off the hook.

Andre Blay and Magnetic Video

Looking at the video industry in the fall of 1976, a Michigan businessman named Andre Blay who had made a career as a distributor of audio and video equipment also saw an opportunity. Blay had recently moved his company, Magnetic Video, into video production and duplication for the corporate market, and he believed that the market for prerecorded tapes could encompass far more than advertising and corporate training videos:

I was confident that there was going to be a hardware war and that there would be an ample number of machines. But I had to fight with almost everybody including my banks, my employees and a lot of other skeptics that those hardware owners would want software. But we were aware of a market called the 8mm market where the studios were licensing to the 8mm manufacturers, 20 minute versions of the movies; it was almost like a synopsis . . . I said, well, if they can make $40 million doing that, we can make a hell of a lot more selling the full version.[38]

With that in mind, Blay sent a letter to the heads of the major Hollywood studios asking for permission to license their movies for distribution on videotape. Today, many in the video industry acknowledge this query as the beginning of the industry as they know it, but in 1976 the studio response was less than welcoming. Most didn't even bother to write back, and only Twentieth Century Fox, which was itself preparing to test the waters of the prerecorded videotape market, responded affirmatively.

Over the next year, Blay and Fox executives hammered out a deal that ultimately required an advance of $300,000 to Fox, plus a minimum of $500,000 per year against a royalty of $7.50 per tape sold. In return, Blay's Magnetic Video would have the nonexclusive right to sell videocassette copies of fifty movies from the Twentieth Century Fox vaults, chosen from a list of 100 titles selected by Fox. From today's point of view, the choices were less than ideal: the most recent movie was four years old, and they had all previously been sold for broadcast on network television. On the other hand, these were full-length, uncut Hollywood movies like *Patton*, *The French Connection*, *M*A*S*H*, and *The Seven Year Itch*, and this was the first time that mainstream consumers could actually buy their own copy to watch whenever they wanted.[39]

So how to get the tapes into the hands of those consumers? Blay tried to package videotapes with VCRs: "We called on every set manufacturer or set importer and tried to strike a deal where we would advertise our titles in return for some price concessions. For instance, we convinced RCA to give away one of our movies to every person who bought a machine in the month of August."[40] At a time when Sony was still advertising the Betamax as a time-shifting device, these original equipment manufacturer (OEM) schemes helped link prerecorded tapes to VCR hardware.

Blay also created a successful mail-order business for his movie catalog through the newly established Video Club of America (a division of Magnetic Video). In 1977, just before the holiday season, he placed an advertisement in *TV Guide*, proclaiming "50 magnificent movies you can show anytime on your own TV."[41] The ad emphasized the personal freedom of owning a videotape copy of a movie:

You now have the opportunity to select and show the original full-length uncut versions of Hollywood's finest movies on your own TV set . . . at less cost than taking your family out to a theatre.

These are the spectacular big-budget features . . . starring Hollywood's most famous stars . . . superbly photographed, most of them in vivid full color.[42]

One thing to notice about this text is that while it uses the same rhetoric of freedom and personal empowerment as the Sony advertisements describing time shifting, it takes the extra step of emphasizing the quality of the available movies. The description of Magnetic Video's movies as "original full-length uncut" sets up an explicit contrast with movies as shown on television, which were edited for content by censors and to make time for commercial breaks, while the references to "spectacular big-budget features" with "famous stars" "in vivid full color" seem to jab at the existing market of older, public domain black-and-white movies.

The Video Club of America advertisement appeared in copies of *TV Guide* nationwide, costing Magnetic Video $65,000 to reach more than 21 million homes, and within weeks more than 12,000 customers had sent in a $10 deposit to become charter members. At the time, however, less than one percent of American homes owned a VCR, and if Blay were simply interested in targeting VCR owners, he might have spent his money more efficiently by advertising in mainstream video publications such as *Video* or *Videography*. The *TV Guide* advertisement does hint, however, that more was at stake than simply signing up VCR owners to buy tapes, since its opening words aren't addressed to VCR owners at all: "If you've been waiting for the finest in home entertainment before investing in a video-cassette unit, your timing is perfect." Moreover, the first item listed as a "preferred benefit" of joining the Video Club of America is the ability to "save $100s off the list price of the newest Sony Betamax and RCA SelectaVision videocassette recorder/players." Magnetic Video wasn't just selling tapes to people who already owned VCRs, they were also selling the very *idea* of movies on videotape as a reason to buy a VCR in the first place.

Though initially popular, only about 10 percent of Magnetic Video's sales came via mail order, and the Video Club of America was quickly eclipsed by in-store sales.[43] Looking back, Blay recalled, "I was taught somewhere in my college education that the early stage of product cycle and the final stage of product cycle is when mail order flourishes. The reason it flourishes at both of these times is that there aren't enough retail stores promoting the product to reach the potential customer base. When we went into the mail order business, we only had 700 stores nationwide carrying our product."[44] In order to grow beyond those 700 stores, Blay initially drew on those with expertise in video—the electronics reps and distributors who connected VCR manufacturers with electronics stores.

The first firm to answer Blay's call was B&S Sales, a New York City–based company consisting of its two owners, Arthur ("Artie") Bach and Bernard Herman, and one salesman.[45] B&S Sales had been representing various video and audio manufacturers in the tristate area for decades, and when Blay placed an ad in the *New York Times* looking for representatives, Bach and Herman quickly jumped at the chance. In August 1977, B&S Sales became the first Magnetic Video reps in New York.

The actual responsibility for introducing stores to the new movies on videotape fell to B&S Sales' lone salesman, Wayne Mogel, who claims to have sold the first Magnetic Video cassette in the country. His first customers were the electronics stores along Fifth Avenue in New York, with their front windows crammed full of cameras and electronic gadgets. At the time, these stores were selling 8mm film cameras and projectors, as well as digested 8mm versions of mainstream movies, and their owners were enthusiastic about Mogel's wares when he came to sell them "Hollywood in a box."[46] He would take their orders, write up a purchase agreement, and send it back to Michigan, where the cassettes were boxed up and shipped directly to the store. In stores themselves, the Magnetic Video tapes were generally sold alongside video recording equipment (costing around $80 apiece, these tapes were expensive, high-prestige merchandise).

Though it's unclear exactly who did so first, was at this point that distributors and retailers started referring to prerecorded tapes as "software."[47] The choice of language is telling—the manufacturers' reps in particular were used to dealing in electronics like stereos and television sets, which were traditionally called hardware. "Software" as a category only exists in opposition to hardware; as Mogel put it, "software was something that went into a machine . . . it wasn't what they would consider your normal hard goods."[48] There was something more to this definition, however, than software simply being something that was inserted into a machine. After all, blank tapes in themselves weren't software—they were simply accessories. In calling prerecorded tapes "software," these electronics distributors and reps were marking out a specific relationship between prerecorded tapes and video recorders that ignored the videocassette's materiality, literally naming it in terms of the software it contained rather than its tangible hardware nature.

Over the next year, prerecorded video took off like a rocket. Blay expanded his catalog, signing deals with Viacom, Avco Embassy, and the estate of Charlie Chaplin, among others.[49] Within a year, there were so many orders pouring in from stores around the country that Magnetic Video simply couldn't handle the volume. Blay decided to switch from a manufacturers' rep model to a distributor/wholesaler model, convincing some of his earliest reps to focus full-time on video software distribution.[50] Their day-to-day practices remained roughly the same, but rather than sending orders off to Michigan, these distributors would now take payments from retailers and fill orders themselves from a stockpile of cassettes stored back at the office.

From Electronics to Music

By 1980, two parallel changes in the manufacture and distribution of movies on videocassette had dramatically changed the landscape of the business. First, the major studios stirred from their relative slumber and started their own home video divisions. Second, the electronics distributors who had handled lines from Magnetic Video and other fledgling producers of prerecorded videotapes found themselves in competition with an increasing number of music distributors, who had decided that videotapes were more promising than LP records. The combination of these two influences reshaped the distribution network into a hybrid of the music and the electronics industries, changing the meaning of movies on videotape in the process.

Andre Blay's success with the initial Magnetic Video line of Twentieth Century Fox movies had broken the ice, and within three years, most major studios had opened their own home video divisions. Warner Brothers Home Video opened its doors in 1978, followed the next year by Paramount, Columbia, and Twentieth Century Fox (which, rather than start a new operation from scratch, simply bought Magnetic Video for the sum of $7 million). Disney Home Video joined the fray in 1980 along with Metro-Goldwyn-Mayer, which established a short-lived partnership with CBS, and they were soon joined by United Artists, Orion, and MCA/Universal (who couldn't deny the growing video market, even though their lawsuit against Sony wouldn't be resolved for another three years).[51]

These home video divisions were primarily responsible for the production and advertising of their studios' movies on videocassette. Especially as a flood of entrepreneurs rushed to open video stores, the studio operations had no more interest in dealing directly with retailers than had Magnetic Video, and they looked to the preexisting distribution network to carry their new products. Many of the early Magnetic Video distributors began to carry lines from other studios as well: B&S Sales, for example, found their video distribution sales growing so rapidly that they dropped their hardware lines and changed their name to Star Video, moving their offices from New York City to New Jersey and opening a full warehouse.[52]

Around 1979, an entirely separate group of distributors began to take notice of the success of prerecorded video. With the entry of the major Hollywood studios into prerecorded videocassettes, many observers in the music business thought that the time was right to move into the burgeoning business of video—the record industry was in a slump, and video seemed just the thing to perk up flagging revenues. Music distributors like Noel Gimbel (one of the first to sign up when Magnetic Video switched from a rep model to a distributor/wholesaler model) understood video through the lens of LP records and audiocassettes. For them, the technology was irrelevant—they were in the business of selling home entertainment.

That's not to say the transition from music to video was always easy. Don Rosenberg was a salesman for Schwartz Brothers, one of the larger record distributors in the mid-Atlantic region, when the company decided to get into the video business: "They just said, 'Look, if you want to switch over, we're going to start this video division.'" When the new video division was ready to go in early 1980, however, Rosenberg wasn't sure where to start:

I remember the first day that I had to call on video stores, and my first thought was, "Where are they? Who am I gonna call on?" I've never seen a video store. I've never heard of a video store, and I know I can go to my record accounts and try to sell them video, but aside from that I don't know where I'm gonna go. . . . I was literally going strip mall to strip mall trying to see, you know, if there are any video stores here.[53]

At the dawn of the video industry, distributors like Rosenberg literally had to hunt for stores to stock their wares, trying to convince existing music retailers to stock both music videos and feature films, while simultaneously scouring the yellow pages and searching shopping malls for new video stores.

Stylistically, the two types of businesses were quite different—many of the first video retailers thought of themselves as part of the electronics industry, dealing with high-end merchandise, and expected their distributors to be clean-cut and polished. At the same time, Rosenberg recalls, "with record stores you'd wear blue jeans and a T-shirt . . . if you showed up dressed too nicely in a record store back then, they were suspicious." Sometimes he had to change clothes between accounts to match his customers' expectations.[54]

The differences between record and video retailers went far beyond clothing; in those days, the music and the electronics industries had two very different ways of operating, based on two very different understandings of the product they were carrying. Electronics distributors were used to selling tangible artifacts—brown goods were *things*, and their distribution was based on the simple idea that when a retailer took delivery of a shipment, he or she paid for it quickly (generally within thirty days). The music industry, on the other hand, didn't see itself as selling *things* so much as *entertainment*. LP records and audiocassettes existed in a strange ontological limbo where their material nature was overshadowed by the songs they carried. This understanding of the nature of their product was reflected in the music industry's way of doing business—"you paid as you sold."[55] In essence, music recordings were treated the same as theatrical movies, with the distributor (and thus the studio) getting paid only when the consumer herself paid for the entertainment. If a particular album didn't sell, the retailer had up to ninety days to return it to the manufacturer, free of charge, an arrangement unheard of in the consumer electronics industry (though quite familiar to another kind of media store, the bookshop).[56]

There were also structural conflicts between the music and the electronics industries. The music industry was organized along geographic lines—a given music distributor was granted an exclusive territory for each of its product lines, in return for which he or she agreed not to sell outside of that territory. The electronics industry, on the other hand, was more of a free-for-all, with competing distributors in various markets cutting deals with anybody they could, regardless of location. Magnetic Video's early reliance on electronics reps seems to have started the video industry along the latter path, and as they began to sell to distributors around the country, they didn't try to set up exclusive deals by region (in a particularly egregious example, two Philadelphia-based Magnetic Video distributors opened up across the street from each other).[57] When music distributors began to carry video, they lobbied studios for exclusive territories, and "the electronics guys, they actually hated the record guys because they didn't want [the video industry] to follow that pattern."[58]

Though neither side entirely prevailed, the distribution network for prerecorded videotapes that emerged assumed a de facto shape that was roughly to the liking of both camps. Initially, studios allowed distributors to sell anywhere in the country and retailers tended to buy prerecorded tapes from whoever had the best price. As more distributors opened their doors, however, the logistics of physically shipping cassettes meant that in order to get new releases as soon as possible, it made the most sense for retailers to buy locally. At the same time, when someone called a studio asking how to buy its movies on tape, "even though there was no exclusive territory, [the studios] still would funnel back to the guy in that territory."[59] Though this compromise tended to satisfy the distributors who were used to the music industry, they were never entirely happy about it—looking back, Don Rosenberg calls the lack of exclusive territories "very short-sighted, very stupid . . . it created a terrible distribution system that plagued the business for years."[60]

Mediators

As a category, mediators such as the distributors I've described have been generally overlooked by historians of technology in favor of the more traditional designer/user schema. Though several scholars have examined the roles of such mediators as home economists and rural sales agents, these actors are more often extensions of the manufacturers (both of consumer goods and, in the case of home economists, knowledge).[61] Though their role in mediating between users and producers requires the flexibility to interact with both groups, such mediators might be more correctly categorized as popularizers rather than free agents, bound (uncomfortably, at times) to the corporate and governmental interests who hire and train them.

The early distributors of prerecorded video, on the other hand, played a much more explicitly creative role, one that focused on establishing new meanings for the VCR rather than mediating the existing meanings between various social groups. They weren't part of a larger strategy on the part of the film industry—much the opposite, they came into video from very different backgrounds and were often at odds with their studio suppliers. When

Andre Blay sent letters to the major film studios, he did so as a hardware distributor, and Jeff Tuckman came up with the idea of selling public domain films on video as a record wholesaler, not as a filmmaker. They invented the industry as they went along rather than following any directives from (or even the will of) the product manufacturers, and often found them-selves in the position of having to sell their ideas both up and down the distribution chain.

In their book *Analog Days*, Trevor Pinch and Frank Trocco tell a similar story about one of the earliest Moog synthesizer salesmen, David Van Koevering.[62] Van Koevering, according to Pinch and Trocco, was a "man with a vision," an ex-evangelist preacher who was "the sort of guy who could sell anything." At the time, the Moog was primarily a studio instrument: Bob Moog (its inventor) and his engineers were well aware that the synthe-sizer was difficult to set up and keep in tune, and the dizzying array of knobs made it difficult if not impossible to replicate a sound exactly during differ-ent performances. Van Koevering, then running a traveling novelty music show explaining the basics of musical instruments to schoolchildren, saw past these limitations and believed that the Moog could be as popular an instrument as the electric guitar. Setting out to make it so, Van Koevering bought one of the smaller Moog synthesizers, wrote "MOOG" on the side that would face the audience, and hit the road, adding the new synthesizer to his show.

After a while, Van Koevering grew tired of his demanding performance schedule and decided to begin selling Moog synthesizers full-time. He found musical instrument retailers initially resistant, sending him away with the caveat that "[I]f you can prove to me that musicians will [use the Moog], you come back . . . and I'll sell them."[63] Since retailers were initially dubious, Van Koevering went over their heads, directly to musicians themselves. Traveling from town to town, Van Koevering would find local clubs and set up a Moog free of charge for the night's performance. Once rock keyboardists had a chance to try the instrument out for themselves (and see the crowd's reac-tion to it), many were hooked. The synthesizers were expensive, but Van Koevering often had loan papers on hand immediately after a performance for the musicians to sign. Traveling from town to town and hiring others to do the same, Van Koevering built what he called "a Moog network selling Moog synthesizers coast to coast" through his company, VAKO (after VAn KOevering).[64]

Though born out of a similar populist impulse, Van Koevering's distribution network, which eventually did count retail stores among its customers, came to a very different end than did early video distribution. When Bob Moog's company was bought out by a venture capitalist, the new owner brought Van Koevering (formerly an independent actor) into the company, threatening to cut off VAKO's supply of synthesizers entirely unless he agreed. Of the early video distributors, only Magnetic Video was formally absorbed by its supplier, in a deal welcomed by Blay (it was bought out in 1979, becoming Twentieth Century Fox Home Video).[65] Thanks to the movie studios' lack of awareness and the diversity of program sources, suppliers couldn't exert the same kind of pressure on video distributors as had been exerted on Van Koevering. By 1980, the increasingly entrenched video distributors were acting as autonomous entities, having staked out their ground between movie studios and retailers without becoming solely beholden to either.

While studios controlled the manufacture and licensing of their movies, video distributors arguably had a larger role in shaping the industry. As we've seen, distributors were responsible for convincing studios to market movies on videocassette to begin with, as well as shaping the networks by which those tapes would reach retailers. For the most part the relationship between distributors and studios was friendly and close, but when they found themselves at odds, the distributors didn't hesitate to cross studio instructions. As for distributors' relationships with retailers, they were close and supportive (as we will see in the next chapter), though also at times fraught with tension.

These distributors were very clear about their identities—they explicitly understood themselves as mediators between producers and consumers, entrepreneurs who had seen an opportunity. Pinch and Trocco draw an analogy between Van Koevering's actions and Latour's analysis of Pasteur's enlistment of new groups in the cause of his anthrax virus. The same analogy extends to the video industry: distributors like Andre Blay and Noel Gimbel had *enrolled* movie studios into a market the studios had initially feared, and established prerecorded videocassettes as more than simply VCR accessories.[66] They defined the video industry's norms, emphasizing the content of prerecorded videocassettes over their high-technology nature, and in the process situating home video somewhere between the consumer electronics and music industries from which it had come.

For most Americans, however, the most visible consequence of the distributors' work was a new kind of store, one that was a cross between a lending library, a brown goods store, and a record shop. These new video stores framed movies on videocassette in a new context and comprised an essential part of the experience of video technology. It is to the construction of these stores, as well as the new kind of mediators who owned and operated them, that the story now turns.

3   Retailers

Alongside the construction of prerecorded videotapes as content rather than technology, retailers began to literally and figuratively construct the spaces in which these movies and other programs might be sold. The distribution network that would bring movies and other prerecorded videocassettes from producers to retailers had been established, but it was by no means obvious how to make that final leap into the hands of consumers. As the video industry initially began to move away from the brown goods stores, it wasn't clear where exactly it was going to end up.

When Ruth Cowan writes about the "consumption junction," she uses the term to refer to the cultural space in which consumers interact with producers, literally choosing among the competing technological frames available within the marketplace.[1] In a market economy, one measure of the success of a consumer technology is its popularity among consumers, who choose a given technological frame (and thus a given technology) by voting with their money—the act of consumption is literally an act of social construction.[2] While Cowan treats the consumption junction as a cultural space, I want to take her well-chosen phrase a step further and discuss the consumption junction not just as an analytic category, but as an explicit physical and cultural space whose geography and norms are manifestations of an underlying technological frame. Thus, I argue, retailers create the retail space as a reflection of their understanding of the technology they are selling, as well as their assumptions about their customers' preferred understanding of the same technology.

In the unstable days before an industry is commonly understood, its retail spaces manifest the various interpretations of the goods inside. As new meanings and frames are created for technologies, they result in varying kinds of stores, and as technologies are brought into preexisting retail spaces, the assumptions and norms of those stores influence still-unformed perceptions of the technologies themselves.

In the case of video, the brown goods stores that offered videocassette recorders along with blank and prerecorded video cassettes in the late 1970s were designed to foster and sell that emerging technology within a particular frame. The fetishization of technical artifacts in these stores—from walls of television sets all showing the same image to microwave ovens displayed on pedestals—isolated these technologies from the context of their use. The organization of a brown goods electronics store tells customers to judge the products within based on their technical features, from the size of the screen to the number of amps in each speaker channel. As the previous two chapters have showed, video was initially situated within the brown goods stores, and many early VCR owners bought into this technocentric videophile understanding of video.

Videocassettes were initially understood as VCR accessories, and as such were displayed alongside VCRs in brown goods stores. Between 1977 and 1980, however, the spaces in which recorders and cassettes were sold began to change—an increasing number of movies and other programs were released on video, and this "software" began to overshadow the hardware. Tape displays soon took up more floor space and attracted more consumer interest than the newest recorders, while consumers were often astonished to find actual *movies* available for sale.

Arthur Morowitz, a New York video distributor who believed that "the retailers who were trying to get into video weren't merchandising properly," took the next logical step. In May 1978, Morowitz opened his first Video Shack store at 49th and Broadway with an inventory of 600 titles (the majority X-rated) and no VCRs. Seeing movies on videotape as a high-end consumer good in their own right, Morowitz constructed his retail space as a hybrid of the brown goods store and the movie theater. The store was decorated with "blinking marquee lights" and movie paraphernalia like "posters, a director's chair . . . a stuffed King Kong," but the videocassettes were displayed under fluorescent lights in glass and chrome jeweler's display cases.[3]

Though explicitly foregrounding software on its own merits and leaving out hardware entirely, the initial norms of Video Shack were very much in the mold of the brown goods ethic. Cassettes were offered for sale only, and the target customers fit "the established demographic patterns of home video buyers: male, over 35, upper middle to high income levels." Morowitz believed that "it pays to have salespeople who are experts," and the store was run by a "neat, happy, agreeable, pleasant, and trained" staff in the vein of brown goods salespeople.[4] The staff worked on commission and was trained in the art of the hard sell: one employee recalled being taught to "put the product in the customer's hands, so they have a sense of ownership and they're more likely to buy."[5] Moreover, the videocassettes were initially grouped by program label rather than by theme or title, in a scheme analogous to the clustering of all Sony electronics or Frigidaire appliances in one brown goods display.[6]

Across the country in southern California, a retailer named George Atkinson was simultaneously creating a very different kind of consumption junction for videocassettes. Atkinson had previously built a business by renting public domain movies on Super 8mm film (and later U-Matic videotape) to hotels and pizza parlors as a form of free entertainment for customers, and he understood the Betamax not as high technology, but simply as the "perfect movie machine."[7] In 1977, when he heard about the initial batch of movies on videotape released by Magnetic Video, Atkinson looked at his own business and figured that the demand for movie rental would be leaps and bounds above the demand for 8mm movies and projectors. He placed an ad in the *Los Angeles Times* that read "Video for rent" along with a coupon for readers to fill out and mail in, and within a week, Atkinson said, "I had about a thousand coupons."

Unfortunately, Atkinson had no inventory, and without investment capital found himself in a catch-22: in order to make money, he needed a library of videocassettes to rent to customers, but he couldn't afford to buy the cassettes without rental revenues. He soon hit on an ingenious solution: renaming his store The Video Station, Atkinson began charging $50 for an annual membership, and quickly raised enough money to build a library of tapes.[8] Once a customer had paid the membership fee, he or she was entitled to rent tapes for only ten dollars per night, and though Atkinson was happy to sell tapes outright to eager customers, he quickly found that rentals comprised the bulk of his business.

In later years, many in the video industry would look at George Atkinson as the father, if not the patron saint, of video rental. While Arthur Morowitz had inverted the traditional foregrounding of hardware over software, George Atkinson had reconceived the relationship between his product and his customers. The Video Shack model framed movies on videocassette as artifacts, tangible objects to be displayed and sold to consumers for their own use. On the other hand, The Video Station offered videocassettes to consumers not as things to be bought, but as experiences to be rented. The former technological frame situated videotapes firmly in the preexisting constellation of consumer electronics, while the latter defined a new space, one more akin to the movie theater or concert hall than the record shop. Atkinson's model ultimately came to define the video store as a cultural institution, though it would take years (and thousands of prospective retailers, each informed by his or her own understanding of movies on videocassette) before the modern video store was a stable fixture in the cultural landscape.

Fotomat

In the late 1970s, studios were still just beginning to follow of Twentieth Century Fox's lead into this unfamiliar new world of video, and the path was by no means well defined. In 1979, for example, executives at Paramount Studios decided to start releasing films on videotape. However, rather emulate Twentieth Century Fox and others who had relied on the developing network of video distributors, Paramount decided to go it alone, and in April announced an exclusive arrangement with Fotomat to rent and sell movies on videotape.[9]

At the time, Fotomat was a film-developing chain with upward of 3500 locations, best known for its small, hut-like booths located in shopping mall parking lots across America. Based on an understanding of video founded not in the entertainment business, but rather in photography, Fotomat executives had decided early on that consumers would flock to video cameras in the same way that they bought 8mm film cameras and 35mm still cameras, mostly using the technology to preserve important events and memories for later viewing. From the Betamax onward, Fotomat stores stocked blank tapes alongside 35mm camera film, and in 1978 the chain began offering a successful film-to-tape service, transferring slide shows and home movies shot on 8mm film onto videocassette.

In addition to a familiarity with video technology, Fotomat locations had another quality that lent itself to the prerecorded video business—customers who dropped off film for processing had to return the following day to pick up their prints. This fit perfectly with Paramount's plans for renting videos to consumers—customers would call a toll-free number a day in advance, then pick up their movies from the nearest Fotomat kiosk when they dropped off a roll of film. For a special introductory rate of just $6.95 per night, they had their choice from a catalog of 130 movies, most from Paramount Studios.[10]

The initial Paramount-Fotomat plan constructed video in the image of the theatrical movie industry. The Paramount Antitrust Case of 1949 had established that studios could not have interests both in film production and exhibition, which left studios in the business of producing movies and distributing prints to theaters, who would then collect a fee for each viewing, passing a percentage of the proceeds back to the studios.[11] The Fotomat arrangement seemed analogous, except that in this case the customer wasn't paying to watch the movie in a theater but rather in her own home. Following Fox's success with Magnetic Video, Paramount offered the tapes for sale as well as rental, but they were pinning their hopes on the latter rather than the former.

In mid-1979, the first Paramount videocassettes were shipped to Fotomat locations in California for a trial run (the program went national in December of that year).[12] The test market was a success, except for one thing—video distributors around the country were frustrated by Paramount's exclusive deal with Fotomat. In particular, retailers and consumers around the country clamored for copies of *The Godfather* and *The Godfather Part II*, two immensely popular Paramount movies available only through Fotomat. Distributors accommodated them as best they could—the most common tactic was to find a friend or business partner in Los Angeles who would run from Fotomat to Fotomat buying copies of movies at $80 apiece, then ship them east to a distributor who would sell each for as much as $150.[13] Even worse from Paramount's perspective, the tapes were often sold to retailers who would then rent them to customers, with Paramount never seeing a penny of those rental fees.

Ultimately, it took a distributor to clean up this mess for Paramount. Noel Gimbel, whose renamed Sound/Video Unlimited was by this point one of the top video distributors in the country, contacted Paramount and asked if they would let him distribute their tapes. He was told that Paramount was quite happy with their experience in California, and when asked why they should deviate from their plan, Gimbel put it bluntly: "Because it's not working."[14] He explained what distributors had been doing, and his description of the underground market in Paramount videocassettes was apparently successful, because later that year Sound/Video Unlimited became the first wholesale distributor of Paramount Home Video's product line.

Fitting Video into Existing Stores

As word of successful experiments like Morowitz's and Atkinson's stores began to filter around the country through newspaper articles and distributor word-of-mouth, retailers in existing industries began to consider a stronger emphasis on video. The most obvious contenders were the brown goods retailers, who had been selling consumer electronics for decades and who had made a place in their stores for VCRs and video technology, only to find that the video was far more profitable than audio. As their video sections expanded, audio technology was crowded out, and many of these retailers decided to focus solely on video (much as Artie Bach and some of the brown goods distributors discussed in the previous chapter had dropped other product lines to specialize in video).

Michael Becker, for example, owned a hi-fi store called The Sound Room on New York City's Upper East Side and in 1977 came across some early Betamax and VHS recorders when he went to his stereo distributor to pick up an order. Figuring that they might appeal to his customers, Becker bought a few machines and started selling them alongside the stereo equipment. "Then, I saw that [the distributor] got in the first group of prerecorded video from Magnetic Video . . . and he had them in a case," Becker recalled. "I said, 'you know, not only do I want the movies, I want to buy the case so I could show them something' . . . in the midst of this stereo store, we started selling videotapes and recorders."[15]

As video distributors and studios increasingly oriented themselves toward prerecorded tapes, traditional home electronics dealers were forced to make a choice—"Stick with hardware or jump into the business of renting prerecorded software?"[16] For Seattle's Magnolia Hi-Fi and Video chain, the vision was of an all-encompassing "home entertainment market": "We feel that audio and video seem to be merging into home entertainment. To do well in audio, you'll have to do well in video, and vice versa, because ultimately they're going to be one and the same."[17] For audio stores that embraced this holistic vision of home entertainment, a move into video often entailed little more than adding "& Video" to the end of their store name.[18] Michael Becker, however, saw the future and profits in video, and aimed to become "a real killer video store."[19] Becker found a partner and slowly phased out audio, renaming his store The Video Room.

Electronics stores weren't the only ones in the music industry to get into video—by 1980, record stores were feeling the pinch of declining sales. Many music retailers tested the waters of video as a way to increase revenues, often finding that "video sales were great, and audio was starting to bottom out."[20] Storeowners were frequently able to order videotapes from their existing record distributors, making it that much easier to ease into the new industry. If they didn't close their record stores altogether, record retailers who dabbled in video often opened up separate video stores, adjacent to or even sharing an entrance with their existing record stores.[21]

Large record store chains also held their metaphorical fingers to the wind and embraced prerecorded video to varying degrees. Camelot Music, for example, had been in the record business since the late 1960s, and found itself under the same financial pressure as smaller record retailers. "Getting into video was a logical progression for us," said camelot's video buyer in 1983. "The record industry had become stagnant, and we found that we were vying with home video for the entertainment dollar." When the first Magnetic Video tapes came out, Camelot dove in headfirst, with as many as 100 of their 140 stores across America selling prerecorded videocassettes by early 1980. When the market shifted from sales to rentals, however, executives were spooked by the alien business model and sold off their inventory, staying out of video for almost two years.[22]

Sometimes the impetus to try video came not from within a record store, but from someone outside it. Jack Messer had been working in his family's Cincinnati construction business for several years when he decided to turn his personal interest in video into a career. He arranged a meeting with executives at Newmark's, a local record chain, suggesting that they open a video department. They agreed in April 1980, so Messer bought "two used bakery cases" and had videocassette orders air-shipped from New York and Florida to Cincinnati, opening his first video department four days later. Within a year, Newmark's video department was outgrossing its record department in sales and was rechristened The Video Store, with Messer in charge. By 1985, he owned eight freestanding video stores, and was the dominant video retailer in Cincinnati.[23]

Still other storeowners found a way into the video business not through consumer electronics or music, but through cameras. Rather than situating video technology in the realm of home audio or television, these retailers saw video as a natural extension of amateur photography, only using magnetic tape instead of film. Video initially appeared in camera stores as an alternative to Super 8mm home movie cameras, and (as discussed in the last chapter) mainstream film-processing outlets like Fotomat began to stock videotapes alongside film.

Some specialty camera stores, however, underwent the same transformation as Michael Becker's Video Room—as the video section expanded, it began to crowd out the more traditional camera business. The Media Center, for example, had first opened in Pittsburgh as a camera store with darkrooms for rent. Owner Henry Tolino and his partners were interested in video but couldn't afford to buy equipment for themselves, so they saw branching out into video as "an opportunity for us to get to play with all the neat equipment."[24] The Media Center continued selling camera equipment, but through appearances on local newscasts and other publicity quickly developed a reputation as the place to go for video in Pittsburgh.

Other retail spaces decided to grab a piece of the video action, most notably convenience and grocery stores. Some convenience stores, particularly those in rural areas, found video rental a natural fit and began to stock prerecorded cassettes behind the counter. Grocery stores soon followed suit.[25] The first to place videotapes in grocery stores were rack jobbers, distributors who would stock and maintain a small rack of products as a sort of a store-within-a-store. One of the earliest rack jobbers, Massachusetts' Super Video, would set up in supermarkets using three-sided kiosks at the end of an aisle that contained baseball-card-sized badges featuring available movies, which customers would exchange for the actual tapes at the service desk.[26]

Eventually, larger grocery store chains like Publix, Safeway, and Kroger's began to develop full video departments within their supermarkets. These departments were often run by the store itself, and tweaked the notion of the video store to the needs of the grocery. In a 1984 interview, the president of grocery chain Von's described his ideal video section in terms of convenience: "In a supermarket, people are in a hurry to get out. They're loaded down with hamburgers and milk, so we've designed a system that gets to the customer quickly."[27] For the businesspeople who ran supermarkets, convenience tended to mean stocking only a limited selection of new releases and children's video; as they saw it "You can go buy a tennis racket at a mass merchant, but a serious tennis player goes to a sporting goods store."[28]

In perhaps the most unexpected framing of movies on video, the truck and trailer rental franchise U-Haul decided to enter the fray, announcing in early 1984 that it would offer video rental in 1,100 of its company-owned stores across the United States and Canada by June of that year. The so-called Haullywood centers planned to each offer a rotating selection of about 100 videos from a vast central library of "foreign films, classics and special interest product in addition to hits." The fascinating thing about this ill-fated U-Haul initiative (the pilot programs failed miserably, and the program was quietly discontinued within months) was that it framed the business of movies on video purely in terms of rental. U-Haul was thus particularly well positioned for the video rental market because, as video rentals manager Karen Fernow explained, "What other company has been in the rental business for more than 40 years?"[29]

Finally, one seemingly natural retail space for movies on videotape was also the most controversial: movie theaters themselves. In a 1981 trade journal article with the blunt title "The Home Video Explosion—Will Exhibitors Let it Pass By?," Arthur Morowitz made the argument that theater owners should embrace the new medium: "Theater owners know the business of entertainment cold. They possess a knowledge that no other group of businessmen has in this country. They know motion pictures and how to sell them. It follows that they are the best prepared group to become videocassette . . . retailers."[30] Another trade journal article exhorted theater owners, "Let's not forget that tapes are movies, and movies as we all know are the heart and soul of the exhibitor's business . . . If someone's going to take money from our pockets, it might as well be us."[31]

A handful of independent theater owners around the country followed this advice, devoting space to a fledgling video section.[32] One such store, Pine Hollow Video, was opened in early 1982 when the son of the owner of Movies at Oyster Bay, a twin cinema in Oyster Bay, New York, began "dealing through the box office window with a total inventory of about 25 tapes." Soon the rental section expanded into the theater's lobby, and by 1984 comprised about two-thirds of the theater's daily revenue with a rental library of "some 3800 tapes."[33]

For the most part, the few exhibitors who successfully branched into video ran independent theaters, usually located in the center of small towns. The one notable attempt to bring video rental into a larger movie theater chain was a deal struck in 1982 between the National Video franchise and the Moyer Theater chain, which owned twelve movie houses across the northwest United States.[34] The theater chain had already created a video store subsidiary when Ron Berger, the head of National Video, proposed a trial period renting of videos in Moyer Theater lobbies. In short, the experiment "failed miserably," according to Berger, because "people rent where it's convenient and theaters tend not to be convenient to where you live, they're five miles away or ten miles away and that's not on your way home from work."[35]

Ultimately, though even movie studio executives tried to convince theater owners that "Selling Videotapes Isn't Selling Out," few exhibitors moved past their fear of the technology into video retail.[36] The one sort of theater that did embrace the new medium with open arms was the adult movie theater, in whose lobbies or adjacent storefronts video became a prominent fixture. In fact, predating the announcement of Video Shack by a good four months, an ad appeared in *Videography* proclaiming "The largest selection of Adult rated video cassettes in New York . . . Sweetheart's Video Centers: in the lobby of the Theatre at 153 W. 49th St . . . we stock Quality X in both Betamax and VHS formats!!"[37]

The Gold Rush

In 1980, less than 3 percent of American homes with television sets owned VCRs. Just four years later this figure was close to 20 percent, and as video became increasingly popular among consumers, many would-be entrepreneurs saw an opportunity.[38] While retailers of many different stripes continued to add video to their preexisting stores (with varying degrees of success), by late 1980 there were roughly 900 stores across the United States that put prerecorded videocassettes front and center.[39] These storeowners didn't necessarily have experience in electronics or music retail—many had no previous experience in retail at all. Arthur Morowitz characterized these early days of the video industry as "a gold rush, this business. Who comes to a gold rush? People who don't have a lot of estate to leave. People who got fifteen thousand bucks from their aunt and their brother-in-law and every friend they could scrape up. People who had no idea what they were getting into."[40]

An independent survey in late 1980 of 214 "retailers who are primarily in the business of selling and renting pre-recorded video cassettes, rather than those who are primarily hard goods dealers," offers a unique snapshot of these emerging retailers and their stores at the beginning of the "gold rush."[41] According to the survey, more than half of such stores had been open for less than nine months, and more than three-quarters were turning a profit. These retailers were scattered across the United States, though particularly concentrated along the West Coast, and almost 60 percent of the stores were located in strip malls (as opposed to those located in enclosed malls, downtown areas, or freestanding buildings, which each comprised 10 to 15 percent of the sample).

Perhaps the most intriguing thing about this survey, however, is the glimpse it offers into the backgrounds of these video retailers. Only a little more than a third of the storeowners surveyed came from a retail background; the majority were professionals of one sort or another who had simply decided to open a store. The list of previous occupations is astonishingly broad, covering a full spectrum of professions from accountants to disk jockeys, real estate agents to plumbers (notably, "housewife" was included as a profession, comprising close to 5 percent of interviewees). This sheer diversity of backgrounds, evident at such an early stage in the industry, raises a question—if these individuals didn't have a retail background, why did they

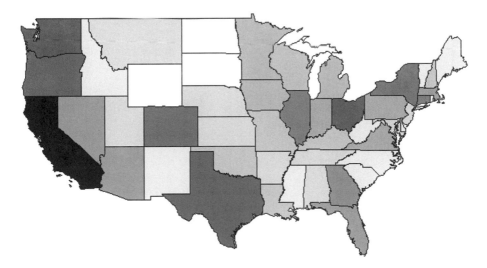

Figure 3.1
Geographic distribution of stores focused on prerecorded videocassette rental/sale as of October–November 1980, with darker colors representing higher numbers of stores per state. (*The 1980 National Video Research Survey*, Video Consultants.)

Previous Occupation	Percentage of Owners Responding
	11.7
Student	
Wholesale/Industrial Sales	1.6
Professional Occupations	51.6
Contractor	6.3
Housewife	4.7
Television Industry	3.1
Marketing	3.1
Certified Public Accountant	2.3
Video Manufacturing	2.3
Real Estate	2.3
Computer Service	1.6
Management Consulting	1.6
Banker	1.6
Insurance Agent	1.6
Wholesale Lumber	1.6
Electricion	1.6
Disk Jockey	1.6
All Other (no more than 1 each)*	16.4
	35.2
Retail Sales	
Video & Television Stores	8.6
Stereo/Hi-Fi Store	4.7
Printing/Office Supply	3.9
Furniture	2.3
Clothing	2.3
Automobile Sales	3.1
Computer Sales	1.6
Telephone Sales	1.6
All Other**	7.0

* Professionals in the "all other" group include doctor, movie theater owner, farmer, pilot, truck driver, photographer, secretary, pharmacist, and plumber.

** Retail salespersons in the "all other" group include butcher, beauty shop operator, liquor store owner, other food occupations, and stained glass making.

Figure 3.2
"What did you do for a living before you opened your video store?" (*The 1980 National Video Research Survey*, Video Consultants, courtesy Ron Berger).

decide to open video stores, and what sort of understandings of video technology informed the stores they opened?

Some retailers wound up in the video business almost by accident, because of a business tip or a family member who had heard something from someone about this new video thing.[42] Others were just in the right place at the right time, like Chicago's Michael Weiss, a production assistant at an audiovisual firm who happened to attend the meeting of the International Tape Disc Association where Andre Blay announced his deal with Twentieth Century Fox. Weiss saw the promise of the idea, scrounged up $30,000, and opened his first That's Entertainment store in November 1976.[43]

In the end, even though there were retailers with experience in electronics, records, and cameras, the vast majority of video storeowners I've interviewed or surveyed shared one common motivation: they were movie buffs. These retailers were usually drawn to the video industry not because of a deep-seated fascination with the technology of Betamaxes and videodisc machines, but because of what these machines could do. The idea of bringing movies into people's homes captivated them, and they loved the idea of owning and working in a store full of movies.

For camera store owner Mike Salomon, moving into video wasn't so much an opportunity to tinker with the technology as it was an opportunity to indulge an interest in movies. Salomon had worked in the motion picture business in New York through the 1960s, first as a messenger and then as a commercial editor, but had moved out to New Jersey in 1972 to open a chain of camera stores with his brother (who had experience in camera sales). Following a 1979 tip from a friend in advertising that "you'd be able to buy movies and watch them in your house," Salomon began to stock movies on video in his Hazlet, New Jersey camera store. Though demand was high, with customers driving up to an hour from Brooklyn and Staten Island, Salomon's employees were dubious, asking him, "Are we going to be a video store or a camera store?" Returning to his love of movies, he decided to emphasize video, and opened several video-only stores over the next few years.[44]

Steve Savage, on the other hand, was already working in the movie industry as an assistant manager at the Bleeker Street Cinema (later to become the Angelika), but he dreamt of something more: "Some people have wanted to or had fantasies of opening up a restaurant; I wanted to open up a movie theater." In 1980, when he read an article about video rental stores opening up in California, Savage realized that video technology could make his dream come true. "Instead of having people come to my movie theater, I would take my movie theater to their homes." Without even owning a VCR himself, he got a few friends together as investors, and New Video opened its doors in Greenwich Village in November of 1981.[45]

Another film buff, Brad Burnside, graduated from Northwestern University with a degree in film, and spent his college years booking 16mm and 35mm films to show on campus. After graduating in 1976, he went to work for an electronics retailer in Evanston, Illinois selling televisions, stereos, and the newest product on the market, the videocassette recorder. He was an early adopter of the technology, becoming "sort of a Mr. Video among the various sales people,"[46] and when the store closed in 1980 he decided to open Evanston's first video specialty store, renting and selling movies on videocassette.

Many new retailers had no formal background in film whatsoever, but found themselves drawn to the business of renting and selling movies for different reasons. Mitch Lowe, for example, who had fled the United States for Europe in the 1970s to avoid the draft, returned to the San Francisco Bay area to find a world of popular culture that he'd missed entirely. Video technology offered a way to catch up: "I was like, you know just a kid in a candy store in the early '80s."[47] He split the cost of a Betamax with his brother and spent a lot of time renting movies from the local Captain Video store, getting to know the management quite well. When the storeowners wanted to expand, Lowe offered them a loan, and was eventually made a partner. Within two years, he realized all the things that he could do better on his own, and opened the first store in his Video Droid chain in 1984.

In his 1991 masters' thesis, titled "The New Nickelodeons", Daniel Moret draws an explicit comparison between the heady days of the video retailing gold rush and the nickelodeons of the early twentieth century. As he explains, the nickelodeon owners who "would rent an empty storefront, put in some chairs, install a projector and screen, and hang out a sign reading 'movies'" seem not too far removed from the aspiring video retailer who "would rent a small storefront, install crude shelves, purchase a few hundred videocassettes, and put out a sign which said 'videos.' "[48] With the barriers to entry low and the potential revenues high, nickelodeons proliferated rapidly, with nearly 20,000 theaters dotting the landscape by the mid-1910s, quickly becoming the most profitable form of mass entertainment in the country before eventually being replaced by a network of more "reputable" theaters.[49]

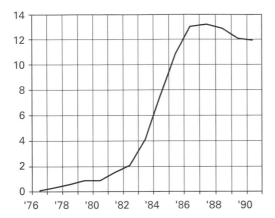

Figure 3.3
U.S. VCR Sales, in millions of units sold. (Moret, *The New Nickelodeons.*)

Video stores followed a similar trajectory, and by 1986 there were at least 25,000 retailers across the United States who, whether driven by a love of movies or simply a desire to cash in on the latest trend, saw tapes rather than VCRs as the main product they were selling.[50] In the independent (also known as "mom and pop") video stores of the early 1980s, prerecorded movies and other materials on videocassette were at the center of the consumption junction, and hardware like recorders was shifted to a subsidiary role. Unlike the earliest days of video, where prerecorded tapes were sold as accessories to play on your VCR, recorders were now explicitly sold as machines that could play movies from the local video store, and it seems no coincidence that sales of VCRs dramatically increased as video rental stores proliferated.

Distributors as Knowledge Brokers

Though they came from widely varied backgrounds, video retailers did have one thing in common—they bought their tapes from the same distributors. Storeowners who had moved into video from other kinds of retail were used to working with distributors, and often their transition into video was eased or even inspired by their existing distribution relationships. Like Michael Becker, many stereo and audio equipment retailers were introduced to video by their hardware reps.

Moreover, as more manufacturers moved into video equipment, distributors found it easier to convince retailers that this new technology was relevant. When Sony and RCA made the only consumer video cameras, for example, it was difficult for camera storeowners to fit the technology into their stores (as they understood them). By the early 1980s, however, traditional camera companies were rebranding other manufacturers' video cameras with their own logos, facilitating the distribution of video cameras through retailers' established relationships with their camera reps. For retailers themselves, "once you could actually buy a Minolta or an Olympus or whatever camcorder or camera, [video] was an easy thing [for camera stores] to take a risk with."[51]

For those novice retailers with no previous business experience, the distributor was far more than just a conduit for products. If the early video industry was a gold rush, these storeowners were the adventurers who hopped a train headed west without knowing the first thing about panning for gold or surviving in the Klondike. In the nineteenth century, such entrepreneurs relied on frontier storeowners, who knew what supplies would be necessary and what would be dead weight. In the early 1980s, however, the general store came to them.

Since one couldn't open a video store without merchandise, a retailer's videotape and hardware distributors often mediated her introduction to the world of video retail. Don Rosenberg, the distributor who initially didn't even know where to *find* video stores in his home territory of Maryland, remembers how the video gold rush started: "People started opening up stores like crazy and I started getting phone calls, because if you called Paramount and said, 'I want to open up a store in Hyattsville, MD,' they would give you our number . . . I was literally working 20 hours a day going to people's houses." Since most of these would-be retailers didn't even have a store location yet, distributors often made house calls: "I would go to somebody's house for breakfast, somebody else's in the middle of the morning, somebody for lunch, and I'd go to somebody's house for dinner. They'd literally serve me. I'd be sitting at their kitchen table eating dinner, writing up an order for movies."[52]

Those conversations over kitchen tables often covered far more than a simple business transaction. Wayne Mogel of Star Video spent countless nights with his clients, going over the ins and outs of running a video store. "[Retailers] relied on their distributor to teach them the business, show them how to merchandise and market their stores—we did everything for them."[53] Once a retailer opened his or her doors, distributors like Mogel and Rosenberg were a continuing source of expertise, giving advice on how to set up and display merchandise, how to advertise, and even how to interact with customers.

For these distributors, many of whom had no explicit experience or training in retail, much of this expertise was built up on the job while visiting one store after another. Like a bee pollinating a field of flowers, these distributors spread information within the network from retailer to retailer, creating a body of shared knowledge.[54] "I got to know what would work in certain stores that had worked in other stores of its like," Mogel remembers.[55] Meeting with his customers on a weekly basis, Mogel developed a nuanced sense of what a video store should be, which he then passed on in his advice to other store-owners.[56] Many retailers recall spending hours every week with their distributors discussing the nuances of their business, such as which tapes they should stock and how many copies (not to mention which format).

Over time video distributors helped to shape the video stores they served, and through the cross-pollination of ideas, they helped to establish some continuity in the industry about what a video store should be. Their influence extended beyond questions of store layout and videotape selection, at times shifting the definition of what home video was and what technologies belonged in a video store. For example, the pairing of VCRs and videotapes with videogame systems and software was facilitated, if not directly spurred, by video distributors who encouraged their retailers to stock machines like the Atari VCS-2600 and Mattel's Intellivision.[57] Distributors saw it as their job to inform their customers: as one said in a 1981 interview, "We have to be able to tell [the retailer] where the business is going; what's available . . . we have to be an expert so that he can be an expert."[58]

Distributors also helped support fledgling video retailers in one other crucial way. In their position as middlemen, distributors were able to finesse the relationship between retailers and movie studios, particularly when it came to money. As discussed in the last chapter, studios demanded payment for videotape copies of their movies within thirty days. Storeowners, however, often relied on rental revenues to be able to pay for the very tapes they were renting to customers. Distributors were integral to the resolution of these conflicting two business models, offering lines of credit to retailers while paying studios up front, in essence allowing both studios and retailers to maintain their incompatible visions of how the industry should be run. By selling tapes to stores on credit, Mogel recalls, "We kept a lot of stores afloat that are thriving now, but were really struggling back then."[59]

Though distributors played an essential role in the early video industry, they found it by no means easy to negotiate the space between manufacturers and retailers. According to Mogel, "We were always caught in the middle, because we had two masters. We had to keep the studios happy, but then again we had to keep the independent retailer happy. It was a juggling act."[60] Successful distributors managed to balance their customers' and suppliers' competing needs, in the process building a shared body of knowledge among the video retailers they serviced.

From the consumer's perspective, of course, none of these layers of mediation were visible—all they saw was the product of these negotiations, the video store itself. Initially, these stores were a diverse bunch (if, in fact, they could be considered stores at all); the initial interpretive flexibility of prerecorded cassettes meant that storeowners presented them in different ways, depending on the technological frame through which they understood their wares. This is not, however, a unique story—the physical space within which *any* technology is sold to consumers is informed by the retailer's technological frame, from a dedicated storefront to a single shelf in the back of a convenience store. From its spatial layout and visual rhetoric to the juxtaposition of some products and not others, the physical space of the consumption junction is a tangible manifestation of the retailer's understanding of the technologies being sold.

Once retailers were brought into a loose network by the distributors who stocked their shelves, the differences between their stores began to level out. By brokering knowledge between storeowners, distributors facilitated the emergence of a common homevideo paradigm, a shared understanding of both what it was good for and how it should be framed for consumers. From the perspective of those consumers in the consumption junction, the relatively stable result of distributors' and retailers' efforts was a particular kind of in-store experience, one that combined the aesthetics and production values of Arthur Morowitz's Video Shack with the videocassette rental plans of George Atkinson's Video Station. As the video store emerged as a broader cultural institution through the 1980s, a new VCR owner's first experience in a video store would come to include the same essential aspects, regardless of the storeowner's background.

4   Movie Culture in the Video Store and at Home

As the video store became more of a phenomenon in its own right, owners increasingly designed their stores to reflect the cultural norms and expectations surrounding the movie industry.[1] Most retailers, schooled by their distributors to see their wares as movies rather than cassettes, built or adapted their stores to embody the technological frame of the VCR, essentially creating a movie theater in the home. While this general theme dated back to the earliest video stores (recall the marquee lights and directors' chairs that decorated Arthur Morowitz's Video Shack), store owners continued to co-opt the rhetoric of the movie theater, at times tweaking it to fit their purposes.

A handful of retailers tried to take matters a step further, building video stores that were exhibition spaces in their own right by adding small video theaters. Burt Tenzer, the founder of a successful video store franchise, developed a "portable mini-theater that will fit inside video stores" which included "cushioned seats, curtains, projection system, audio system, lights and concession stand" and required around 400 square feet of free floor space.[2] The owners of a similar facility argued that their store was "just as good as most suburban multiplex theaters with their tiny screens and paper thin walls," and believed that customers would welcome the chance to watch a video out of the home for the same reasons that people go out to eat.[3] The universal lack of success of these initiatives, however, indicates that while consumers understood video technology as being bound up with movie-watching, the act of watching *at home* was integral to this frame.

Thus, retailers adopted the physical rhetoric of the movie theater only on a superficial level—put simply, many video stores began to look like movie theaters. A 1984 trade publication profile of La Mar II Video Movies began: "While some movie theaters are talking about selling videocassettes in their lobbies, one innovative retailer has put a movie theater lobby in his video store." The store, named after a local movie theater that had recently been torn down, was a faithful recreation of a movie theater lobby, from "working arcade games to a popcorn machine and candy concession." After walking through a turnstile into the store, customers might well have mistaken the space for a theater lobby were it not for the shelves of cassettes lining the walls (which they could approach by navigating "theater-style crowd control ropes").[4]

While most retailers didn't take things quite so far, one particular aspect of movie theaters was increasingly common in video stores: popcorn. As one industry insider wrote, "It's so automatic and instinctive to have popcorn while watching a movie that it's almost 'unnatural' not to. It's been shown that even when people buy popcorn to eat at home, three-fourths of it ends up being consumed in front of the TV set. What a natural tie-in to the home video customer shopping for movies in a video store."[5]

Popcorn itself had been a fixture in the American movie theater since the 1930s, when theater owners looking for an additional revenue source began to open concession stands in their lobbies (beforehand, popcorn, candy, and other similar products were sold in shops adjacent to or wagons parked outside of theaters), and it carried a strong association with movies as a cultural institution.[6] As theater attendance declined, worried popcorn manufacturers spent millions of dollars to market popcorn as a snack for home consumption, with brands like TV Time and Jiffy Pop offering a quick and easy-to-make accompaniment for home television viewing.[7]

Though a popular at-home snack by the 1970s, popcorn still retained a close association with the movie theater, and was specifically sold in video stores as an accompaniment for watching movies, not television. A 1987 advertisement for Orville Redenbacher "video popcorn" made this pitch explicit with the proposal "How's this for a great double feature! My Gourmet Popcorn and your video movies" alongside a doctored photo of Orville himself standing amidst the Marx Brothers, a juxtaposition that reinforced "video movies" as being like real movies, while at the same time offering Orville Redenbacher's popcorn as the preferred brand for in-home consumption.[8]

In addition to selling popcorn as an accessory to movies on video (in much the same way as videocassettes were originally sold as accessories for VCRs), many retailers used popcorn as a design element in their stores. One retailer owned an "old-fashioned popcorn machine" and gave customers free popcorn every time they rented a movie, using as much as fifty pounds

Figure 4.1
Advertisement for Orville Redenbacher Popcorn. (*Video Store*, 1987.)

of corn on a holiday weekend, with the explanation that "[the popcorn] creates a movie theater atmosphere," while the owner of the eponymous Bob Bardash Video went so far as to actually install a fan to blow the aroma of fresh-popped popcorn out into the street to lure customers.[9]

This attitude wasn't limited to popcorn; a spokesperson for Nestle Foods said of its attempts to market Raisinettes, Goobers, and other movie-theater staples in video stores that "The theater experience is exactly what we hope to duplicate when we merchandise our candy products through video stores."[10] Retailers agreed: "We're no longer settling for being seen as just a video store. We're creating an atmosphere, an entertaining environment and candy is just a part of that aura. It makes people feel good."[11] This atmosphere stretched the bounds of traditional movie theater rhetoric in some surprising ways—for example, several retailers discovered that it could be extremely profitable to offer pizza alongside their popcorn and candy, offering consumers a night's dinner *and* entertainment in one fell swoop.[12]

Rental vs. Sellthrough

Along with the creation of a video store that emphasized movies rather than hardware came the retailers' question of how best to make money from their software inventory. Among many in the industry, an understanding of movies on videocassette as experiences to be rented rather than objects to be purchased was simply taken for granted. As a Fotomat executive said in 1980, "With most movies, people don't want to see them more than once. Owning them is only for snob appeal and I think that appeal will diminish after they've invested a certain amount of money in cassette movies."[13] Surveys and sales figures seemed to bear that opinion out, with one researcher later reporting that nearly nine out of ten VCR owners had rented a prerecorded cassette in the previous year, while less than a quarter had purchased one.[14] Conventional wisdom aside, it wasn't clear to many retailers whether this consumer preference was a reflection of a deep understanding of movie-watching as an ephemeral experience, or simply the fact that even though prerecorded videocassettes had been taken down from their pedestal as high-tech consumer goods, their prices remained at those lofty heights.

According to the "first sale" doctrine, retailers who purchased a movie on videotape and rented it to customers were not legally required to give the movie's makers a cut of their rental profits. In the early 1980s, various studios had instituted rental-only plans in an attempt to get a piece of these profits, but their efforts were a dismal failure thanks to a retailer revolt (which we will get to in chapter 6). Thus, from the studios' perspective, the only way to make money on home video was to charge prices that reflected an estimate of the given cassette's potential rental revenue, leaving the price of the average mainstream movie on videocassette upward of $79.95. While these high prices seemed relatively reasonable to retailers, who could count on renting a tape countless times (and perhaps even reselling it to a customer as previously viewed), they were prohibitively expensive for most VCR owners, especially as VCR sales increased beyond the enthusiast markets.

A handful of movies *had* proven themselves so popular that they would sell at virtually any price, most notably *Star Wars*, which was released on videotape by Twentieth Century Fox in May 1982. Thanks to George Lucas's and theater operators' concerns about videotape sales interfering with the film's theatrical re-release (timed to lay the groundwork for the premiere of *Return of the Jedi* later that year), the *Star Wars* videocassette was to be rental-only for its first three months of availability. Consumers demanded the ability to buy the title, however, and dealers were more than happy to oblige, selling copies at up to $120 each under schemes such as "three month trials" or even "lifetime rentals" that technically adhered to Fox's requirements.[15]

A few months later, Paramount decided to take a chance and offer *Star Trek II: The Wrath of Khan* on videocassette at the unheard-of price of $39.95. It cannot be overemphasized how risky this was for the studio, and how it represented a significant trust in retailers' good intentions. "Before dealers decide merely to take some quick profits with what will no doubt be a strong rental tape, a few facts should be mentioned . . . this is not just a Paramount test, but a test for the whole business," wrote an editorialist in *Video Business* at the time. "In order for the test to be a success, Paramount must sell significantly more cassettes of *Khan* than it has of anything else . . . if the test succeeds, dealers will in all likelihood see a greater number of 'A' titles from all the studios at low price points. We don't have to tell you what a shot in the arm this could be for your business. On the other hand, if sell-through does not increase dramatically . . . it is possible the studios have not yet begun to fight on the first sale doctrine."[16]

As it turned out, the *Khan* test was a rousing success, wildly outpacing any previous sales records, and Paramount struck pay dirt again with the release of *An Officer and a Gentleman* in February and *Flashdance* later that year, each for $39.95.[17] Even though other studios followed Paramount's lead, only the biggest hits were generally priced for sell-through, and the industry quickly developed the two-tiered pricing system described by Frederick Wasser: "Videocassettes are either list priced at the high $90-plus or at low prices ranging from the break-even price of $8 through the $30 range . . . The rental stores are free to buy low priced tapes and consumers are free to buy high-priced ones. However, it is generally the case that only rental stores will invest so much money in the high-priced tapes and will have less interest in the low-priced tapes, since part of the audience for such tapes will have already purchased them and will not be renting these titles."[18]

With rental firmly established as the dominant frame for understanding most prerecorded cassettes within the video store, the space of the video store was constructed to fit this paradigm. Rows of videocassette cases dwarfed the few floor models of recorders and cameras, and the overwhelming impression was of a sea of movies and other content rather than the technological fetishism of the brown goods store. Pumping revenues back into their rental libraries to keep up with new releases and customer demands, storeowners found themselves on top of an ever-growing inventory of cassettes. Shelves lined store walls, and when video retailers found themselves out of display space, they simply added more freestanding shelf units.

Rental cassettes themselves, however, were rarely placed on the shelves, at least at first—at around $80 each and small enough to fit into a purse or coat pocket, videotapes were far too valuable to leave out in the open. Rather than going with such a live inventory system, video store owners usually filled their display shelves with empty boxes and kept the cassettes themselves behind the cashier's counter or in a back room. The resulting configuration offered customers the ability to physically browse through a store's inventory without the risk of shoplifting. As one store manager remembers, "You [could] spend a length of a movie trying to find something with your friends . . . You had fun. It was like a little event."[19]

By keeping the actual rental cassettes behind the counter, accessible only to employees, video stores helped alienate the content recorded on the tapes from their material nature. In the video store, customers browsed *movies*, represented by boxes containing nothing more tangible than the experience of watching a movie itself. The boxes, usually featuring still images as well as a short description of the movie or other content, served as placeholders—

Figure 4.2
Videocassettes shelved behind the counter at The Video Droid, a video store north of San Francisco. (Photo by author.)

signifiers referring not to a specific cassette, but to the idealized text encoded therein. While most stores retained at least the suggestion of videotape as a material artifact in the shape of the boxes lining their walls, a handful took this division even further—at Steve Savage's New Video, videocassette boxes were flattened and laminated onto "browsers" for customers to flip through like LP records.[20]

This tension between physical cassette and idealized content, exemplified by dual sets of shelves (one for customers, the other for employees), was a distinct problem for video retailers, as the connection between the two wasn't always obvious. As inventories expanded, it became more difficult for clerks to quickly find a requested tape. The constant influx of new inventory meant that it was extremely labor-intensive to keep the actual cassettes alphabetized for easy retrieval, and many retailers created their own numbering schemes to match cassettes to boxes, though these numbering systems meant that it was even more difficult to find a requested cassette by name (unless the customer found the matching box first).[21]

In order to imply as wide a selection as possible, early video retailers often left empty boxes on the shelves even when the corresponding cassettes were not available, meaning that consumers often selected box only to find that the corresponding videocassette had already been rented to another customer. Moreover, as inventories grew, retailers often displayed only one box for a given movie, even if multiple copies were available in either the same or different formats. To resolve these dilemmas, retailers created a range of tokens to represent each physical instance of a given movie, from small laminated cards to plastic discs attached to the front of display boxes with Velcro or hung from hooks alongside tapes.[22] One customer remembers quite clearly the frustration of finding an available tape in the appropriate format at the local store,: "[there were] 2 hooks under each box [with] little chips on each hook for copies in stock. Yellow for Beta and Red for VHS. Nothing was SO disheartening as going in to get that film you wanted and seeing only a Yellow chip, you just wanted to scream out—FUCK BETA!"[23] When the token signifying an available copy wasn't there, the empty box taunted customers with the knowledge that however much the film existed in an idealized sense, the physical limitations of videocassettes meant that they wouldn't be watching at home it that night.

While this in-store ecology of boxes, tabs, labels, and chips provided an interface of sorts for the video store's available inventory, in practice video rental required a moment of substitution in which the *idea* of a given movie was exchanged for an actual *cassette*. It was only in the act of paying for the rental that the movie took corporeal form, as the brightly-colored display boxes or tokens were exchanged for anonymous plastic Armory cases holding prerecorded videocassettes.[24] It's ironic that at the moment the movie became tangible for the consumer, a unique token representing the movie was substituted for a physical object that, with the exception of a label on the cassette and perhaps another line or two of writing on the case itself, was completely generic, visually indistinguishable from the thousands of other cassettes shelved behind the cash register.

Hardware Rental

This casting of movies on videotape as experiences rather than commodities influenced not just the spatial geography of the video store, but also the material nature of the VCR itself. In order to watch videocassettes at home one had to have a video recorder, and this requirement anchored the experience of the movie in a piece of hardware that permanently resided alongside the television set. The natural chronology of "buy a recorder, then rent movies" made sense when prerecorded cassette viewing was just one use of the VCR. However, retailers found that increasing numbers of consumers were lured into home video mainly by the promise of movies at home rather than any interest in the technology itself.

Previously, the VCR was understood as a technological artifact in itself, while the cassette existed solely as a physical manifestation of a program. However, as one commentator wrote in 1983, "People are starting to realize that owning a VCR doesn't really give them anything but time shift. Most people use the machine to watch programs; there isn't really that much recording being done when you look at the usage figures overall. Sure, you might decide to tape a TV show or a movie now and then, but how many times are you going to watch the same episode of *The Dukes of Hazzard*?"[25]

Along these lines, if the VCR was reconceived by some as nothing more than a transport mechanism for bringing a rented program to a home television set, then it was really no more special than the cassette itself. Many retailers thus began to see the VCR as something that might be rented along with tapes rather than anchored in the home. At first, retailers seeking to expand their customer base were all too happy to rent or lease machines to their customers, until they quickly found that users often had difficulty wiring the machines to their home televisions and, far worse, treated rental VCRs scarcely better than they treated rental cars. The rental machines were constantly in need of repair, and on occasion barely made it out the door: "A customer was taking a machine out as I was coming in. As I started to talk to the owner, we heard a crash from outside. The customer had tried to balance the VCR on his knee while opening his car door and it fell. End of story."[26] Even worse, rental VCRs were a prime target for fraud and outright theft, to the point that one retailer felt the need to fingerprint every customer who rented a machine.[27]

Coming to the retailer's rescue, a handful of companies began to redesign the VCR into a material form that would be portable, rugged, user-proof, and less desirable to thieves. Perhaps the most well-known of these was Superscope's "RentaBeta," which was rolled out with great fanfare in early 1982. The RentaBeta was essentially a seventeen-pound black plastic case "similar in size to a small typewriter" that contained a video player and a cassette chosen by the user, available for the low price of up to $7.95 per night (not including the cost of the tape rental).[28] The cassette itself was loaded and sealed inside the RentaBeta at the video store, and it was a point of pride for a company spokesman to point out that "the consumer will not have access to hardware rentals as a way to get his Beta tapes out the door." In fact, hardware rentals rendered the entire format war between VHS and Beta irrelevant: "The customer who rents the machine doesn't care what format it is, he only cares whether you have movies to show on it."[29]

Figure 4.3
Trade advertisements for portable videocassette players. (*Video Business*, September 1982.)

In response to video store owners' desire to be able to rent more than one title at a time, later machines allowed customers to insert and eject cassettes on their own. Also, while some manufacturers continued to argue that "the customer who wants to record should buy a VCR," others began to include recording capabilities under the presumption that "customers can record special shows . . . for their own use at a later date. Besides, recordability helps blank tape sales."[30]

Ultimately, hardware rentals fell victim to falling VCR prices, and the demand for the occasional RentaBeta or PortaVideo declined. The machines did linger a while longer, however, pressed into service in the seamy

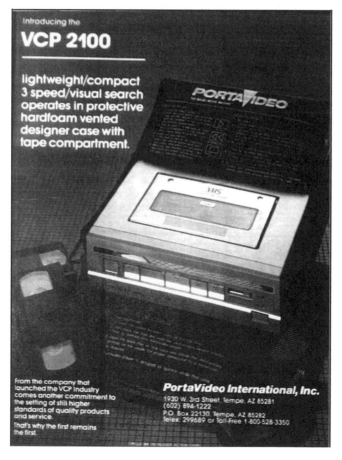

Figure 4.3
Continued

underbelly of video culture. Many users rented a VCR for the night in order to dub copies of rented or borrowed tapes, wiring the machine to their home recorder.[31] Meanwhile, rented VCRs remained a fixture at the ever-popular bachelor party, a late-twentieth-century version of the rented projector and stag film that had been a fixture in American culture for most of the century.[32]

The Theater and the Home

It's important to remember that motion picture technologies involve more than purely technical artifacts, and are embedded in particular social contexts. By the time Americans started to buy VCRs, the theater and the television represented two very different contexts for entertainment: one public, one domestic. The television had replaced the hearth as the focal point of American living rooms, and families who a century before might have gathered around the fire now gathered around their television sets.[33] While many early advertisements offered images of television as a small-scale theater in the home, filled with family, friends, and neighbors, the medium was soon placed in the alternate lineage of the piano and other traditional domestic (and thus private) entertainments.[34] The rhetoric of the "home theater" did persist, but only referring to the relative size and fidelity of its technical components (rather than the social nature of the theater experience).

Once video was a legitimate alternative to the movie theater, some users began to describe their movie-going experience as a function of the theatrical context, rather than simply a consequence of the movies themselves. "A lot of times funny movies are just funnier in a theater with lots of people laughing around you," said one VCR owner in 1985. "We saw [*Caddyshack*] in the movies and we laughed hysterically, everyone did. Then we told friends of ours to rent it, and they watched the whole thing and didn't crack a smile. Some movies just call for getting out in a crowd."[35] Another chuckled over a particular theater experience: "When I saw *Broadway Danny Rose* there was this older woman sitting behind me, you know the kind with the big pocketbook . . . Well, in one scene, Woody Allen is eating in the Carnegie Deli and this lady turns to her friend and says 'I still think the corned beef is better at the Stage.' How can you overhear something like that in your living room?"[36] These sorts of comments were only possible once there was an alternate movie-viewing context that was directly comparable, yet distinct from, the theater.

To many video renters, however, the joys of watching a movie in one's own home far outweighed any romanticized benefits of going out to the theater. They understood the shared context of the movie theater as irrelevant to the movie itself, whose images and sounds were reproduced well enough on their home television sets. Video's domestic context promised the ability to watch a movie free of the hassles of public life, and the popular image of a video viewer was of someone curled up on a comfortable sofa with a pint of ice cream and mound of pillows, happily pausing the movie every time the phone rang or he needed to go to the bathroom.[37] Moreover, video offered freedom from the social norms of the theater; one VCR owner enjoyed movies on video because "you can make all sorts of snide comments without fear of insulting an eight-foot person sitting behind you."[38]

As watching movies on video became a more common practice, the norms of home viewing began to blur into the theater. A wave of complaints surged through the media as film critics (the arbiters of proper theater behavior) led an assault on what Gene Siskel called, "people who chatter through movies, children who race up and down the aisles and the slobs in the seats." [39] These critics pointed their collective finger at movies on video, blaming them for the worsening manners of theater audiences. "Much of the blame for blabbering in the theater belongs to the videocassette recorder, which obliterated the line between watching a video at home and a movie in a theater," wrote one commentator. "Attention spans are shot. The visual and verbal data emanating from the screen does not mean enough to many moviegoers to encourage their undivided attention."[40] There being few metrics for rowdiness in the theater, the actual relationship between theater audience behavior and home video is difficult to pin down, but it's not a difficult leap to connect the idea of the theater and the home as equally valid places to watch a movie with the homogenization of the distinct norms of both spaces.

In addition to viewers' practices, the change in context from the theater to the home had consequences for movies themselves, as the traditional norms of the domestic sphere fostered certain genres that hadn't been as successful in the public space of the theater. The growth in these genres was manifested in the layout of an average video store—while the middle of the store comprised cassettes filed in categories such as Drama, Comedy, or Action, two of the busiest sections were at either end: the children's videos in a front corner of the store, and the adult videos in a back room.

While the Saturday matinee at a local theater had been a staple of many children's lives through the 1950s and 1960s, the theatrical demand for movies targeted directly for children had been waning when video arrived on the scene. For parents increasingly struggling to balance the demands of family and career, video's convenience proved an irresistible technological aid to childrearing.[41] "Parents like to have videocassettes around all the time, so that they can put them on if their kids get cranky," said Duke Kreps, Michael Becker's Video Room partner, in 1982. "A lot of people are calling them electronic babysitters."[42] Hardware manufacturers were just as aware of the marketing potential of the genre—a promotional photograph for the RCA videodisc shows four children gathered wordlessly while watching a Muppet video, their real-life Kermit and Miss Piggy puppets cast aside.[43] The rhetoric of the VCR-as-babysitter saturated the discourse about children's video, or "KidVid," and was only heightened by video producers' discovery that children would watch favorite videos dozens of times without losing

Figure 4.4
Promotional still for the RCA videodisc player. Note the Kermit and Miss Piggy puppets lying on the floor, discarded by the rapt children in favor of their televised counterparts. (Collections of the David Sarnoff Library.)

interest, making parents far more likely to buy children's videos outright than any other genre of prerecorded videocassette.[44] By 1985, children's video had surged to 10 percent of video rentals and sales, growing faster than any other genre, and video suppliers responded with an "onslaught" of new programs produced specifically for the video market.[45]

Pornography, while an established genre long before the advent of home video, was also given a huge boost by the shift from the theater to the home. As we have seen, adult videos were some of the first prerecorded videocassettes available for sale, and production increased dramatically through the early 1980s (by 1984, the number of new adult titles to reach the market each year had risen from around 400 to 1,700).[46] While many of these new video producers offered traditional adult films, some pushed the boundaries of the genre, casting their videocassettes as "men's video magazines" in the mold of more respectable publications like Playboy (which itself began offering "playmates without staples" on video in 1982).[47] Adult film was a natural fit for the domestic space since that (according to conventional wisdom) was where most sex took place. "My wife hates porno films," said Al Goldstein, the publisher of *Screw* magazine and the producer of the popular "Adult Blue" series of adult videos. "She says the rare one she sees that succeeds in turning her on, leaves her frustrated because what can we do in a movie theater? But in your home, it permits you to follow the logical progression of making love with your wife or girlfriend . . . it's just more natural at home."[48] Moreover, the increasing adoption of home video cameras allowed individuals with exhibitionist tendencies to create their own amateur pornography that, if the filmmaker was so inclined, occasionally wound up on the shelves of the local video store for others to rent.[49]

KidVid and adult video were strongly associated with activities traditionally reserved for the domestic sphere (childrearing and sex, respectively), and thus occupied an awkward position within video stores that had been constructed to frame video in terms of the theater. For obvious logistical reasons, the children's and the adult sections were usually kept at opposite ends of the store, but the two areas were both constructed as quarantines. The children's section served as a babysitter, a space where parents could leave children while they browsed the rest of the store. The kids' section was often stocked with toys, coloring books, and lower, child-height shelves. It was designed to keep the rest of the store out, while the "back room" was structured to keep a certain genre contained within its swinging door or beaded curtain. In this spatial layout, the video store reinforced the identity of children's and adult video as distinct and separate from the more theatrical genres.

Ultimately, the framing of the VCR as a medium for movies called into question both the idealized nature of those movies as well as their broader cultural context. Negotiations over the ideal format for a movie, from the shape and size of the screen to the social context in which it was viewed, opened up larger questions about the nature of movies themselves. Over the 1980s, a stable understanding of the relative merits of the movie projector and the VCR emerged, based on a sophisticated yet rarely discussed understanding of the relative benefits of the two technological systems. Movies that traded on their epic nature (particularly big-budget Hollywood blockbusters) were most popular in movie theaters, where their loud explosions and awesome visuals could be reproduced in the most spectacular fashion. As for smaller, character-driven films—particularly as the cost of theater tickets rose to unprecedented heights—they were perhaps best characterized by the oft-heard disclaimer: "I'll wait for that one on video."

5   Retailers, Employees, and Consumers

If a VCR owner from the year 2005 found herself transported back to a video store circa 1982, she would likely find the space remarkably familiar: videocassette cases lining the walls, a movie playing on a television set, maybe some marquee lights framing an announcement board over the checkout counter—all in all, not terribly unlike the video store of two and a half decades later.

However, understanding this consumption junction simply as a transactional space in which money is exchanged for goods and services misses the forest for the trees. That same time-traveling video renter might find herself startled when the employee behind the counter asked her if he could recommend a particular movie, or by the knot of customers hanging out by the counter, shooting the breeze and talking movies. The early video store was often a place to talk as well as to shop, and the persona and expertise of retailers and clerks structured the consumer experience of home video as much as the shelves on the wall or the movie theater trappings. The people behind the counter, and in many cases those in front of it, were not merely moving through the consumption junction, but were in fact integral parts of it.

Consider the English pub, the French café, the German beer garden, or the American tavern—these are places where patrons can buy food or drink, but an analysis that only focuses on the producer/consumer exchange would overlook the vital role these and other spaces play in the establishment and maintenance of broader social relationships. Sociologist Ray Oldenburg describes such spaces as "third places," a category that he uses to refer to "a great variety of public places that host the regular, voluntary, informal, and happily anticipated gatherings of individuals beyond the realms of home and work."[1] Such third places serve the function of uniting the neighborhood through simple, day-to-day interaction. For example, Oldenburg

writes, "In many [American] communities, the post office served this function well when everyone had a mailbox there; when everybody had to walk
or drive to it; and it was kept open, by law twenty-four hours a day. Though
there was no seating, it was a place where people met and conversed, at least
briefly, with each other."[2]

Third places offer spaces within which communities can cohere. A local
bar, for example, offers its patrons a place to socialize outside of home and
work, establishing a sense of commonality between its denizens. The primary mechanism for this community building is simple conversation—
according to Oldenburg, "Nothing more clearly indicates a third place than
that the talk there is good; that it is lively, scintillating, colorful, and engaging."[3] This orientation toward lively conversation is a social leveler; third
places "counter the tendency to be restrictive in the enjoyment of others by
being open to all and by laying emphasis on qualities not confined to status
distinctions current in the society. Within third places, the charm and flavor
of one's personality, irrespective of his or her station in life, is what counts."
Another core attribute of a third place is the sense of shared ownership felt
by its regulars, regardless of who actually possesses the deed to the building.
"Those who claim a third place typically refer to it in the first person posses-
sive ("Rudy's is our hangout"), and they behave there much as if they did
own the place."[4]

For much of American history, the theater served as such a third place. In
his social history of American audiences, Richard Butsch charts the important position of theaters at the center of public life. In early American life,
the theater was a space in which classes mingled, and which served as a
nucleus for "community conversation and civic participation."[5] Though
Butsch argues that theaters shifted from public spaces to private spaces
focused on "shopping and consumption," it is clear from his narrative that
the theater continued to serve as a gathering space, albeit one with an admission charge. Describing late-nineteenth-century Yiddish theater, for example, Butsch writes that "the theater was a social center. The Lower East Side
provided little public space other than the streets [and] theaters were among
the few places where people could gather at little cost; they were commercial
substitutes for the piazzas and other places in European villages and towns,
where these immigrants had been accustomed to gather and talk."[6]

The appearance of motion pictures on the media landscape fit well with
this preexisting function of theaters in cultural life.[7] Nickelodeons' lower

prices earned them the nickname "democracy's theater[s]," and an even more diverse set of patrons mixed within their dark rooms.[8] At the same time, in rural America, small town residents had developed a thriving tradition of gathering at local town halls and opera houses for "homegrown and family centered" entertainment such as "performances by local bands, neighborhood amateur singing, recitals, pageants, tableaux, lectures, and political speeches," and the initial appearance of traveling motion picture shows was constructed in the mold of this moral, community-centered vision of entertainment.[9]

Over the twentieth century, Butsch argues, the movie theater receded in importance as a social and community space. By the end of the 1930s, "the movie, not the place . . . [was] the attraction."[10] Talking and sociability were frowned on at mainstream theaters, with the exception of children's matinees, which continued to serve as community events for decades. Meanwhile, through the 1950s and 1960s, the drive-in theater took on much of the role that the earlier movie theater had played, serving as a third place for local youth and families where the space and interactions with other audience members were as important as the movie (if not more so).[11] This drive-in culture began to disappear within a few decades, however, and by the 1970s movie-going had waned as a fundamentally social (as opposed to individual) institution.

The traditional story told about home video is that its coming fragmented the movie-going audience into individuals. Even though they were no longer centers of social interaction, Butsch and others have argued, theaters still brought people together into a shared space and experience, while home video helped to wall them off into their separate living rooms. What this argument misses, however, is the shared practice of *acquiring* those tapes that were then watched at home—the video store became a third place in which customers, retailers, and clerks met, talked, and shared a common experience of movies on videocassette.

Clubs and Newsletters

On the most straightforward level, customers were often quite literally members of the club. As I discussed earlier, many early video store owners found that the easiest way to quickly raise the capital required to build a rental library (and hence to attract customers) was to charge a membership fee. In the earliest days of video, before it was even clear that videotape rental would be legal, some stores created a particular sort of membership: prospective members were required to purchase a videocassette outright for the then-going rate of up to $100, which they could then take home and watch. When the customer returned her cassette to the store, she "exchanged" the purchased tape for another and paid a minor "restocking" fee in the range of $5 to $10. An analogous scheme, particularly common among hardware retailers, involved customers purchasing shares in a video lending library from which they could borrow one or more tapes at a time, each for a small additional "handling" charge.[12]

Looking back, however, most of the retailers with whom I spoke remembered fondly the days when customers lined up to join straightforward video rental clubs, paying outright for the right to pay for movie rentals. At the time, it seemed that the membership fees rolled in whether the retailers were prepared for them or not. Steve Savage, the film buff who cofounded Manhattan's New Video, remembers giving a potential member a sales pitch on his first day in business, rambling on about store specials, the wide selection of tapes available, even his hope of running film festivals out of the store. When the customer agreed, Savage found himself at a loss: "We had no membership applications; we didn't know what to do. Finally, I took out a pen and a paper and I said 'So what's your name,' and I tried to figure out how to go about renting to somebody."[13]

To lure members, some retailers included a certain number of free movie rentals as a bonus for signing up. As an employee at Palo Alto's Midtown Video recalls, these freebies helped to accentuate the movie-going atmosphere: "We gave out actual tickets like you would get at a carnival [or a movie theater], with our Midtown Video stamp on the back of each ticket. We would give out 10 of these tickets [with each membership]."[14] Stores also frequently offered members the ability to prepay for blocks of movie rentals at a discount, giving the retailer money up front to purchase the very movies that might be rented in the future, usually framing the prepaid rentals either in terms of tickets or as a form of money—at New Video, they were called "Video Checks" and were so popular that customers bought up to 100 at a time.[15] At times—particularly as stores began to computerize their operations in the mid-1980s—these prepaid rentals had no physical manifestation at all, existing solely as notations on a customer's account. "The advantage for the customers (beside the free movie) was that they didn't even have to pay, they just came in and it got checked off on their account."[16]

While membership clubs and prepaid rentals were particularly useful for raising much-needed capital in a video store's first months, they also served another, more subtle, function. Jacques Ellul argues that modern propaganda works not just by bombarding an individual with information, but by *enrolling* that individual through her actions; every time a citizen stands at a rally or participates in a meeting, she further commits herself to the given belief system.[17] Though a very different context, the same principle was at play in video store memberships. Actively joining a given video store's club meant that you were more than an anonymous consumer—you were (literally) a card-carrying member of that store. As it was prohibitively expensive to join multiple store's membership clubs, consumers tended to rent from one store at a time even if there were several in the neighborhood (especially if they had committed themselves even further by purchasing prepaid rentals)[18] Thus, economic, geographic, and cultural factors conspired to heighten customers' identification with their chosen video store.

Once a member, a customer's tie to a given video store was often maintained by a steady flow of communication from the store, most notably the ubiquitous store newsletter. The heart of these newsletters was a list of new releases intended to keep customers apprised of the store's inventory (and, owners hoped, lure them in for a few rentals), but they also often included other material that fostered a sense of community, from a Letters to the Editor page to a "Video Classifieds" section that might offer "the means to exchange or sell various video items."[19] While newsletters were advertisements for video store products, they were also intended to "keep the excitement alive for the proud owners of the newfangled machines called VCRs, and they were an inducement to 'join the club.'"[20]

Meanwhile, if a customer stayed away for too long, store newsletters served as the primary reminder of what she was missing, though a storeowner might occasionally resort to more drastic tactics to reel long-distant customers back to the store. Bob Caras, the owner of a Video Biz store in Washington, DC, sent fake bills for $1,000 to customers who hadn't visited the store in more than ninety days, offering them a free rental when they came charging into the store waving the bill. Echoing the ribbing and joking described by Oldenburg among third place regulars, Caras explained, "We only do it to customers who we know can take a joke, and they're flattered that we think they can."[21]

In rare cases, the newsletter and other publications were in fact the *only* connection customers had with their video store. In 1979, Michael Becker's Video Room began offering home delivery to his customers on Manhattan's Upper East Side, and up to half of his customers rented movies entirely by delivery.[22] Customers relied on a catalog of the store's tape library, as well as a monthly update listing new releases, in order to make their selections.[23] Other stores offering home delivery also relied on printed materials to offer a simulacrum of the actual video store to customers who wanted to experience it from home.[24]

Mom and Pop

There was no direct precedent for the video store, a place where media texts were rented to consumers on a fee-for-use basis, and somebody had to educate consumers about the norms of this new consumption junction. Thus customers' initial experiences with video were directly mediated by the retailer, the man or woman behind the counter who lived a life immersed in the new technology of video. As one retailer recalls, "After a while you'd train the customer, you'd say 'Look, if there's nothing behind the front box, that means it's out,' because we've put all of our cases behind the box."[25] Another had a "whole little tour" which she would offer to each new customer, explaining both the layout of the store and the particular system by which tags hanging next to boxes represented copies of movies available to rent.[26]

The entrepreneurs who opened the earliest video stores were often referred to in both the popular and trade press as "mom and pop" retailers, and while many storeowners loathed the term, it pointed to a fact of life in the average early-1980s video store: retailers and their families staffed the counter, day and night. Many stores were run by a husband-and-wife team, either taking shifts at the counter or dividing up the work by expertise. When they opened their first Video Station store north of San Francisco, for example, Peggy Dorrance (a former librarian) worked the counter and handled the ordering while her husband Ken used his business background to manage the store finances.[27] When the moms and pops had to be out of town, their families often filled in, to the point that the earliest video trade shows were scheduled for the summer, for the explicit reason that it was easier for individual storeowners to travel when their children were out of school and available to mind the store.

Even though the video store was constructed to foreground movies rather than technology, customers who were new to the technology (and had not internalized this particular understanding of the video store) often assumed that their video retailer would be able to offer technical advice and help. Jack Stein, who opened a store south of Los Angeles, remembers potential customers who had already purchased a VCR elsewhere walking into his store and asking, "I got this machine, what do I do with it?"[28] Stein would explain the basics, at times even drawing a diagram explaining how to hook up the recorder. Over time, Stein (who had no formal technical background) moved into basic VCR repair and tape duplication, finding in particular that he could enhance his bottom line by offering basic head cleaning service to his customers. Even though the operation might only require a screwdriver, rubbing alcohol, and some cotton balls, customers tended to prefer bringing in their machines to literally opening their black boxes themselves.

Some retailers, particularly those who were video enthusiasts themselves, were able to provide even higher levels of technical expertise. Michael Dark, an early videophile who helped run several stores in the early 1980s, remembers going that extra mile for his customers: "We'd sell them their second machine and dubbing cables and try to educate them, you know, about making copies. Because we made a lot more money in rental than we ever did on sale of the product . . . we could sell somebody another machine and make a profit there, sell them the accessories, sell them the blank tape. The lowest profit margin of all was selling them the actual film."[29] At times this education went beyond diagrams, with Dark even making house calls to help set up and repair his customers' machines in their homes.

While video retailers offered varying degrees of technical expertise, they discovered that expertise about the movies themselves was even more essential to their business. Thanks to the high initial costs of prerecorded tapes, retailers simply couldn't afford to keep many duplicate copies of movies in stock, so the busier the store, the more likely that the most popular titles were out on a given day. "In the early days," remembers Steve Savage of Manhattan's New Video, "it was really hard to get your first choice when you came into the store."[30]

Retailers quickly found that in order to keep customers happy and money flowing into the cash register, they had to be able to offer alternate movie suggestions to customers, based both on their individual tastes and the simple question of which tapes were available at that moment. "I tried to make psychological profiles of my customers," explains Michael Dark, "to be able to recommend something that wouldn't turn them off." At the same time, this explicit mediation between consumers and a store's video library offered retailers an added bonus: "By knowing your customer, you know what to buy for the next month."[31]

The Video Store Clerk

As stores grew and retailers expanded into multiple branches, it became increasingly difficult for storeowners and their long-suffering families to cover every shift. From early on, storeowners found that they needed to hire additional help, and they looked for employees who would be able to maintain a similar level of customer interaction: "Just being able to be an extrovert, being able to bring up a conversation, be able to look at a customer and know that they're wandering about something and approach them and ask them what they want and what they're interested in."[32]

Moreover, retailers looked for employees who would have the movie expertise necessary to "move the library." A knowledgeable clerk was able to suggest other movies that might appeal to a given customer, combining an expertise in movies with personal knowledge of the customer's tastes, and many stores built their identity on the expertise not just of their owner, but of their clerks.[33] The owners of such stores made sure that their employees could "talk film" with the most discerning of customers, and they looked for this sort of expertise when hiring: "people who've seen a lot of movies and people who have good communications skills . . . people who are actually really passionate about movies."[34]

For stores offering home movie delivery, this sort of employee expertise was essential, because a customer might call and ask, "I've seen *Annie Hall*, can you tell me three other Woody Allen movies that I'd like to see this weekend?"[35] A profile of one video store offering delivery, Philadelphia's Express Video, emphasized the importance of those who answered the phones "transcending the role of order taker and earning a reputation as a friendly and reliable source on the latest and greatest on videocassette."[36]

To test for such expertise, some stores gave employees written or verbal film knowledge quizzes before hiring them, with questions ranging from

"What is your favorite movie, and why?" to "Name three Godard films."[37] "What's your favorite Star Wars movie and why?" was Video Droid's Chris Ritter's favorite question to ask prospective clerks because "in [Marin County], if you haven't seen the Star Wars films, you're going to be a little lost. The only wrong answer is, I would never watch those. As long as you can give me an answer and back up your logic, then you've got the right answer."[38]

The ideal clerk usually came with his movie expertise already in place, and it should come as no surprise that many of the earliest video clerks were film students—New Video, for example, was renowned for employing most of the film students at NYU at one point or another. Most clerks, however, especially those working in cities outside of the traditional Los Angeles and New York film axis, were self-taught. Movie fans across the country found that their encyclopedic knowledge of film was an asset in the brave new world of video rental, and essentially leveraged their pastime into a job. "I was a movie fanatic," remembers Mark Stencel, whose high-school job was at a local video store in Virginia. "I had done nothing for years but sit and watch the movie channel every afternoon after school and had seen probably every major movie that had come out in that period . . . I was very good at finding people's taste."[39]

Frequently a passion for and knowledge about movies was the single most important qualification for the job, even more so than previous retail experience. "We hired somebody [who] had never worked retail," explains Chris Ritter. "He didn't even know how to put a phone on hold. He is a first year film student [whose] great-grandfather or grandfather was a projectionist in Mexico back when you had to hand crank. He has a real love of films and we got very excited to hire him partly because we knew that we could give him as much as he could give us . . . I can teach you to answer a phone. I can teach you to total a bag. I can't teach you to love movies."[40]

Often, clerks were recruited from the customer base, "promoted" in a sense because of their extensive knowledge of or passion for movies. Jerry Frebowitz's first employee at Movies Unlimited was Irv Slifkin, a local college student and customer whose "movie knowledge was amazing."[41] Ken and Peggy Dorrance hired a neighborhood high-schooler who "knew films, loved films," and, finding that it ran in her family, later hired her mother as well.[42] Sometimes persistence paid off: one customer liked her video store environment so much that she kept bringing the owner baked goods and asking for a job. "Finally, there was an open position . . . and she worked [there] for a long time."[43]

Once hired, employers often took explicit pains to cultivate and develop movie expertise in their employees. The most-cited perk of working in a video store, "the idea of having free access to as many movies as I wanted and [getting] paid to sit in a store all afternoon and watch movies,"[44] was as much an asset for the retailer as for the employee. Allowing employees to take unrented movies home cost the retailer nothing, and ensured a staff of happy and informed clerks. Some stores, particularly those which built an identity based on their employees' expertise, went even further; New Video employees were entitled to reimbursements for up to four theatrical movie tickets per month, and if a clerk couldn't afford a VCR of his or her own, the store would loan them the money to buy a machine so that they could build their movie knowledge at home.[45]

The clerk's expertise stretched beyond the borders of mainstream Hollywood film into another important genre for the early video industry: pornography. Paul Fishbein, a clerk at Movies Unlimited, remembers that of "all these new people getting VCRs, the one thing that was common amongst them was, 'Hey, can you recommend an adult film?'"[46] Most Americans at the time had far less experience with adult films than with mainstream films, and many early video store customers looked to the resident expert, the clerk behind the counter. "It was funny," says Michael Dark, "because I was never much into adult films except for the tradability of them, but after a while I'd have to start watching [them] for their quality just because someone's going to ask me for a recommendation, and I'm going to see that customer again next weekend and they're going to say 'You're right' or 'Wrong' or 'Don't ever give me anything like that again.'"[47]

The result of this deliberate emphasis by retailers on movies (rather than on technology) was a particular kind of expert identity for the video clerk. "We prided ourselves on our movie knowledge," remembers Ray Pennisi, a clerk at a Video Library store in southern California.[48] This knowledge was honed and kept visible in countless local rituals and arguments. "They used to argue," recalls Jerry Frebowitz of his employees. "Not *argue*. 'Fun' argue."[49] Video clerks argued playfully: whether a given film was noir or not noir, which was the best *Star Wars* movie, countless bull sessions about genres, characters, directors, and writers.[50] In some stores, clerks found even more public ways to establish their expertise. Steve Savage recalls one particular ritual from his first New Video store: "This was New York, and Martin Scorsese was a big hero to all the film students who had gone to NYU. We had all these NYU film students, and they would put on *Raging Bull*, turn down the volume and recite the dialogue verbatim for the customers. The customers would just hang out at the store and watch it."[51]

Video stores became known as repositories of film knowledge, and at times video clerks found themselves in the role of reference librarians. For example, video stores became an educational resource for local film students, who no longer had to wait for on-campus screenings of classic films. "We get a lot of university students from Berkeley, from Hayward," says Ken Dorrance. "They come all over to see our catalog and our classic section, which is huge."[52] At the same time, local media professionals began to rely on video stores for their film knowledge. "You know, Sinatra died," remembers The Video Room's Michael Becker, "so [local television station] Fox called up and asked 'could you pull together the thirty eight Sinatra movies' and we delivered them to them."[53]

"We used to do a lot of [work] with ILM and Pixar," recalls Chris Ritter. "They would come over and they would pick our brains. 'I need a movie where somebody vomits forcefully,' you know, or 'I need a movie where fire is only on rooftops' . . . They would come to us saying 'what movies can I take back,' and then adapt it into whatever they're doing." One Video Droid clerk was particularly good at this sort of work: "Ing was brilliant. I mean encyclopedic knowledge. Always adding to it, an incredible artist and sculptor too." Eventually, Ritter needed to curtail the outside advice "because they were taking us away from our work," but she tried to convince the effects companies to hire Ing outright, even offering incentives: "If you will give him a job interview . . . I'll wipe out $250 of late fees." She never managed to talk to anyone with the authority to set up an interview, and was frustrated by the experience. "He's got all the information that they want, that they're coming to us for, and . . . because of the way they're organized, [they] couldn't take advantage of it."[54]

For many employees, their time in the video store was "a job, not a career," and it's not surprising that quite a few clerks were interested in making their own movies.[55] According to Michael Becker, "The video store was the restaurant for out of work actors and writers . . . Most of the people that we hire[d] are writers, directors, want to be actors . . . mostly they're not career retail people." By the early 1990s, however, a video store job took on an additional meaning: it was an education in itself. The constant discussion with other clerks, the out-of-work socializing (which often also revolved around film discussions), and the never-ending stream of movies on video made for a thorough curriculum, which dovetailed nicely with the emerging "indie" film movement in American cinema. Thanks in large part to the extraordinary success of writer/director Quentin Tarantino, the video store as a cultural institution became known as the film school of the independent film world.

In 1984, a young Quentin Tarantino started working at Video Archives, a small video store in a Los Angeles suburb.[56] While the store's furnishings might have been modest, its clerks were anything but: the store was a textbook case, with a staff comprised of hard-core customers who had demonstrated their film knowledge and who debated movies endlessly. Though the canonical story is that working in the video store was an education in film history for Tarantino, he tells it a bit differently: "I was already a film aficionado and somewhat of an expert in my own way. That's why I got hired! It wasn't so much that I learned from what I was watching, because I'd been watching the same amount of films anyway. But it was the fact that I got to be almost the Pauline Kael of that video store. I talked about movies all day . . . What we were doing all day, when we weren't helping customers, was hold forth and have heated debates."[57] To Tarantino, it wasn't his expertise itself so much as the day-to-day *practice* of said expertise that is the legacy of his half-decade behind the video store counter.

Tarantino's mythos grew with the mid-1990s success of *Reservoir Dogs* and *Pulp Fiction* and the ensuing press coverage, and virtually every profile or review offered his days as a video clerk as a sort of "street cred" that validated his work outside of the mainstream currents of both Hollywood and film school. "We were the video store generation," said Roger Avary, another Video Archives employee who was Tarantino's writing partner and went on to a successful writing and directing career in his own right. "Right after the film-school generation, the first generation of people who wanted to be filmmakers who had grown up alongside computers, videos, the information highway."[58] Others followed suit; one profile of writer/director Kevin Smith following his breakout festival hit *Clerks* described the new wave of indie filmmakers "emerging out of the video outlets and 7-Eleven stores, weaned on the films that have surrounded them in their time behind the counters of America."[59]

Aside from actual filmmaking, quite a few video clerks parlayed their expertise into future jobs, from film reviewing and other writing to radio or television work. Paul Fishbein of Philadelphia's Movies Unlimited was particularly successful; Fishbein found that so many of his customers were asking for adult film recommendations that he and his coworker Irv Slifkin began publishing a newsletter—"a consumer report type thing" for adult video that he sent to a handful of subscribers and gave away in the store for free. Toward the end of 1983, Fishbein realized that many of his subscribers were fellow retailers, and he decided to refocus *Adult Video News* as a trade publication. "Instead of saying that this is a hot sexy movie that you should rent, [we decided to say] this is a hot sexy movie that you should stock in your store." The magazine was the only trade magazine of its kind, and developed into the primary independent conduit of information between adult film studios and video retailers.[60]

"Like a bar, without the alcohol"

Writing about American taverns, Oldenburg describes the curious position of the bartender: "Good bartenders have the knack of getting their customers together and of making sure that the return patron will have at least one personal greeting each time he or she stops in." The bartender is the host of the third place, "that font of local information, that symbol of authority, that arbiter of disputes, that 'character'," and a good host is essential to establishing and maintaining the vibrant culture of a third place.[61] A key aspect of the video store's clerk/customer relationship was exactly the sort of local authority that Oldenburg mentions—as we've seen, just as bartenders were expected to be up on the concerns of the community of regulars, video clerks were expected to have a certain level of expertise on the items at the center of video store culture, the movies themselves.

In video stores, the man or woman behind the counter set the tone. Many early storeowners were simply social people, eager to engage with their customers. Matt Ratto, whose father Gary owned a video store in Merced, California, recalls one thing above all else: "He just talked. He always talked so much . . . He knew everybody by name, and when people would come in he'd give them recommendations and they'd chat about the weather, what was going on in Merced or whatever." At an acting conservatory years later, Ratto found that a colleague who had lived in Merced remembered his father's store and shared the same memory: "I remember your dad with his big mustache, always talking."[62]

Though early video stores were often seen as questionable influences (thanks to their selection of adult videos), retailers saw themselves as part of a larger community. Gary Ratto, for example, "felt like part of his responsibility was to be able to advise people about what they were going to see, so that they could make judgment calls about whether or not their children should watch it . . . it was very much a neighborhood thing."[63] Retailers were often strong supporters of community activities, sponsoring little league baseball teams and even running voter registration drives.[64] In some cases, video stores reached out to local celebrities; Mark Stencel remembers his boss giving free memberships to members of the professional football team whose training camp was a short drive from the store. For regular customers, it gave the video store an added draw, because "there was a good chance that anytime you came in, in the late afternoon, you could sort of meet one of the Washington Redskins."[65]

As the video clerk's expert identity developed, customer interactions often became less balanced, with clerks asserting their dominance over the movies that formed the video store's core, and thus over the dynamics of the store itself. "Almost all employees were very knowledgeable movie geeks," remembers one Movies Unlimited customer. " They'd talk to you if you were renting something interesting, recommend edgy arthouse stuff."[66] Perhaps the most interesting thing about this comment is the implication that the customer *wanted* to be judged interesting and edgy, cool enough to be accepted by the video store employees. The store was a meritocracy of sorts, and a customer could be welcomed into the inner circle by displaying the same sort of expertise for which the clerks were hired in the first place.

Occasionally employees developed inside jokes that reinforced their identity as superior to customers. Notes were left in customer files, particularly as stores began to computerize their records. The clerk culture was strongly adolescent and strongly male. "I remember my assistant manager friend and I would memorize the account numbers of all the good-looking girls that would come in and rent. We'd say to one another, 'Didn't 57803 look hot last night?' 'Hell yeah!' "[67] Though expert, clerks weren't always trustworthy mediators, and occasionally developed small games that played off the advice they gave to customers. "The most memorable thing to me," recalls an employee of the Stop and Shop Video Center in Milford, Massachusetts, "was on Saturday nights after all the new releases were rented out, having contests trying to rent the worst movies to the customers. I won one night by getting one person to rent *Ishtar, The Sex O'clock News* and *Flesh Gordon* at once."[68]

Video store employees played out this power differential by asserting their dominance over customers. "We knew this customer," remembers Rich Nathanson. "He came in and he's looking at the movies, and he goes 'Santa Claus movie, what's that about?' And the clerk said, 'it's about the fucking Easter Bunny, what do you think?'"[69] Of course, Nathanson is careful to explain, these sorts of comments were made with a wink and the ultimate message that "we're just joking with you," not unlike Bob Caras' fake $1,000 bills sent to customers who could "take it."

While the clerk-customer relationship was at moments tinged with chest-thumping, clerks were usually looked to as benevolent figures who offered good advice and good conversation for those who were looking for either. If anything, customers sometimes tended to overestimate the abilities of the clerk. "The most frequently used in-joke in the store," recalls Pat Nestor, an employee at the Video Quest store in Nestor, New York, "was the 'WITGiTIHS' customers. 'WITGiTIHS' stands for the customers who would come in and ask 'What's In That's Good, That I Haven't Seen?' like we were mind readers and knew what the hell they had already seen."[70] At the same time, those customers who saw the video store as their third place often developed a personal relationship with their video clerk. "[Customers] would come in and they would trust the clerks, and come Christmastime, they used to bring gifts . . . they used to fill up the backroom with customers bringing them bottles of liquor, etc. I remember one [clerk] got a leather jacket."[71] Michael Dark was invited to dinner at some of his customers' homes, and even dated several of his customers' daughters.[72]

While clerks were usually seen as the in-store experts, by no means was the flow of information unidirectional. Many of the earliest video store customers were film buffs themselves, and the lucky clerk who happened to be working when one came into the store might be in for an education. "I remember [one customer who] really knew the old stuff, like the old serials, like old B movies and serials," says Lance Strate, who worked at Arthur Morowitz's Video Shack in the early 1980s. "That's where I first learned about Tom Mix and the Phantom Empire . . . Also, Adolph Greene came in once. I didn't know who he was, and I remember him pointing out *Singing in the Rain* to me."[73]

At their best, video clerks and storeowners fostered a true third place atmosphere in their stores. "There was always conversation in our store," recalls Jerry Frebowitz, owner of Philadelphia's *Movies Unlimited*. "Especially on Saturdays . . . there was so much film talk going on. Total strangers were talking to each other; it was really a phenomenon."[74] Though his customers always came to rent a movie, Frebowitz says, "They definitely stayed to talk . . . hardly anybody came in to get the movie and leave." In other cases, the video store served as a third place for a particular subculture—Shawn Bowles recalls the local video store in Titusville, Pennsylvania: "It was an after school hangout for a few Nintendo, horror film freaks (like me). It was the main piece to our little town's puzzle."[75] Other times the social activity in the store might simply boil down to the clerk and one or two regulars. "There was also an officer who was responsible for patrolling the shopping center who spent most of his patrol hanging out in our store with us watching movies."[76]

One of the most extreme variations on this theme was New Video, which store manager Rich Nathanson remembers as "anywhere from amiable to a jungle."[77] As Steve Savage recalls, "somebody would come into the store on a Saturday night and people were just talking to each other, telling [each other] 'No, no, you don't want to see that' or 'Well if you like that you've got to see this.' Then there would be joking around, people negotiating, 'I'll take this and I'll give that to you.'"[78] Nathanson remembers a similar camaraderie among customers. "People would return bags [of tapes] and we'd open them up. [Other customers] would go 'What's in them, what's in them?'" Savage's fondest memories of New Video were of its energetic and social crowds: "It was the 80s, it was the time that clubs and the nightlife was very, very big with Studio 54 and all these places. We were a part of the circuit. [Club-goers] would come to rent videos before they went out at night because when they got home at four o'clock in the morning, they were too wired to go to sleep and there was nothing on TV, so they would watch a New Video movie. We were the first stop on the nightlife circuit in New York."[79]

Describing his store in a 1984 interview, Savage explicitly described New Video as a sort of gathering place for film lovers. "There are few businesses where people come in two or three times a week. It's more like a local bar— what can we serve you up today?"[80] Bob Caras echoed these sentiments, saying "I almost feel like we're bartenders . . . it's a place for people to go when they want to get away from the office."[81] Along these lines, Chris Ritter of

Figure 5.1
Scenes from the opening night party at New Video's uptown location; the store was a nightlife hotspot. (Stills from a live recording of the event courtesy of Rich Nathanson.)

northern California's Video Droid described her store as "A gathering place, almost like a café without the coffee."[82]

Rhetoric like this points to the status of the early video store as a unique type of consumption junction—one where social interactions were at least part of the attraction for customers. While the raison d'etre of a consumption junction is, by definition, to enable the exchange of goods and/or services between producers and consumers, this social space can take on added layers of meaning and importance. Just as the users of a technology may create alternate uses and meanings for a given artifact, the inhabitants of a consumption junction are able to use it for other purposes in addition to simple consumption. After they opened the store, got hired, or joined the club, many early video store owners, clerks, and customers found that the video store offered far more than just movies on video, and their conversation and camaraderie flourished. By the late 1980s, however, this sort of freewheeling sociability was crowded out by the growing dominance of a handful of video store franchises that reoriented the video store back to its original focus on consumption.

6   Building Closure Around Video (Stores)

One of the biggest questions in any history of technology is how the cacophony of different uses and meanings for a new technology settles down to one stable technological frame. In canonical studies of bicycles, Bakelite, light bulbs, automobiles, and other technologies, scholars have described a trajectory from the flourishing interpretive flexibility of a new technology to the eventual (albeit sometimes temporary) black-boxing of a specific understanding of its meaning and use.[1]

If you accept the general tenets of the social construction of technology, particularly the argument that knowledge of what a technology is good for and how it should be used is distinct from its artifactual nature, then this closure of controversy is ultimately a result of social consensus rather than objective truth. To put it bluntly, a given technological frame that is black-boxed around an artifact is only seen as real and natural so long as everyone involved agrees that it is so. Thus follows a focus on relevant social groups, since the establishment of closure relies on a shared consensus by everyone who is relevant (that is, everyone with power to influence how the technology is understood). Since closure is essentially just a shared agreement on a specific technological frame, once those relevant social groups are all on the same page (either by redefining the technology or by redefining the problems over which they disagreed), closure is achieved. Perhaps just as important, though more overlooked, is the fact that a stable closure is *maintained* only through continued consensus—should the relevant actors and social groups shift their understanding of the technology, the controversy over its meaning can open right back up.

In the case of consumer technologies, Cowan has argued that the notion of the "relevant social group" (which was initially drawn from social studies of scientific knowledge production), can become "potentially infinitely confusing" thanks to the vast array of engineers, manufacturers, designers, managers, distributors, and retailers involved, not to mention the myriad consumers themselves, with their shifting and overlapping identities. Rather than trying to map out this universe of actors, Cowan situates her analysis from the perspective of the consumer, "imagining that consumer as a person embedded in a network of social relations that limits and controls the technological choices that she or he is capable of making." By focusing her historical account on the consumption junction where the consumer makes a choice between competing technologies (or the competing technological frames of a particular technology), Cowan describes this network from the consumer's perspective.[2]

In the early 1980s, that consumer's perspective on home video could be awfully confusing. VCRs had started to receive attention in the press,[3] but for a consumer who wasn't an enthusiast and didn't know anything more about the technology than one might learn from newspaper articles and advertisements, the place to learn more about video technology was the new store with a "VIDEO HERE" banner that had recently opened up a few miles away. Since storeowners came from many different backgrounds, it's no wonder that early customer experiences might vary from store to store; as discussed in chapter 3, some stores presented the video recorder as a high-tech component for the television set, while others described video in terms of cameras and photography, and still others focused solely on the movies you could watch. Depending on which store she walked into, a customer might see merchandise like television sets, photo developers, video games, computers, cables and electrical accessories, books, magazines, and pizza, or there might be nothing but hastily built wooden shelves stocked with as many prerecorded videotapes as the owner could afford.

Thus far, we've seen how closure was achieved on a relatively local level—thanks to the role of distributors as knowledge brokers, retailers who shared a distributor grew to share a common understanding of home video that emphasized content over the technology itself. However, the decentralized nature of distributor mediation meant that while there might be local consensus over the norms and practices of video stores (and thus of video technology itself), there were still disparities from distributor to distributor, or region to region. Within a decade, however, a nationwide closure had been achieved by new institutions that brought coherence to the loose-knit collection of video stores dotting the country, replacing the earlier grass-roots social networks with centralized, top-down structures.

Conventions and Trade Shows

As the number of video stores grew and home video began to develop from a handful of isolated stores into a nationwide industry, retailers quite simply began to talk to one another. In such a new industry it was easy to notice when someone else in town opened up a store, and retailers would hear news about other storeowners from the distributors they shared. This kind of indirect communication led to informal local gatherings where retailers would get together and discuss their businesses directly.

Depending on the personalities involved, the relationships between retailers in the same area ranged from competitive to downright friendly. Frank Barnako, the founder of the local Potomac Video chain, remembers the early market in Washington, DC: "Oh God, we were competitive. You know, sometimes I think the guys took retail spaces to prevent another guy from getting it. We were very, very competitive because I guess none of us knew how big the market was. Occasionally we would get together informally to talk about the business, but these were very stolid conversations because nobody wanted to give away any information or what we thought were secrets. Generally, we'd probably just trade release dates."[4]

The majority of retailers with whom I've spoken, however, seem to remember the period around 1980 as surprisingly easygoing. "You had a few slime balls here and there, but most of your competition was good people," recalls Michael Dark about the northern Californian market. "I've never seen competition getting along so well. If we got burned by somebody, we'd call our neighboring stores and tell them and they'd do the same to us . . . Money was just falling out of people's pockets about that time and everybody wanted a little bit of it. Most of the time, everybody in this business had a little respect for each other, everybody had their own little niche and at that time there wasn't a video store on every corner."[5] In some cases, the friendships forged between storeowners lasted decades. In New York, where there was plenty of business to go around, The Video Room's Michael Becker would meet up with New Video's Steve Savage without any wariness; "We were in different marketplaces, but in the same city."[6]

Initially, the only opportunity for these early video retailers to meet storeowners from other parts of the country was the Consumer Electronics Show (CES), a trade show held by the Consumer Electronics Association in Chicago every summer and in Las Vegas every winter. Though the sprawling show floor covered every imaginable kind of brown good, video technology became an increasingly prominent presence through the late 1970s and early 1980s, with all the major hardware manufacturers featuring their latest models.[7] For video software retailers, however, the excitement was often to be found on the fringes of the show floor, where distributors and program suppliers began to set up booths. "Embryonic home video ventures," wrote a contemporary commentator, "were shoehorned in among watches and jewelry, sometimes sharing space with more established (but still minor) corporate affiliates . . . the experience was closer to a Cairo bazaar than a bona fide industry."[8] Noel Gimbel remembers two things about his first CES show as a video distributor: his location in the very back of the show area, and the long lines stretching down the aisle from his booth of show attendees waiting to meet adult film star Seka, whose tapes Gimbel's Sound/Video Unlimited was distributing.[9]

Building Video Trade Associations

Whether at home or in the aisles of the CES, the initial topics of conversation among early retailers were the day-to-day mechanics of running a video store—what new tapes were coming out, how to set up displays, and stories about customers. Soon, however, bigger issues began to cast their shadow over the video industry. After the appellate court ruled in favor of Universal Studios in 1981, *Universal v. Sony* was on its way to the Supreme Court, and if that decision were upheld some feared that the video industry might collapse. As a possible solution, Senator Dennis DeConcini and Congressman Stan Parris immediately introduced legislation that would allow private, noncommercial recording of copyrighted works in the home, but an amendment proposed several months later by Senator Charles Mathias Jr. threatened to establish a legal principle that video taping in general was copyright infringement (the amendment also established a royalty on each VCR and blank videotape sold, in order to compensate copyright holders for the presumed infringement that blank tapes represented).[10]

At the same time, the major Hollywood studios had begun to take issue with the structure of the rental market. Thanks to the right of first sale, once a retailer purchased a copy of a movie, it was legally her property, and she could rent the tape as many times as she wanted without being obligated to pay the studio a penny more. Needless to say, the studios wanted a piece of this rental action. In the first years of the decade, several studios (including Warner Brothers, Disney, and MGM) offered some if not all of their videocassettes for rental only, requiring varying levels of revenue sharing from retailers.[11] Moreover, Jack Valenti, the head of the Motion Picture Association of America, was beginning to lobby for the repeal of first sale altogether, and some feared that provisions in the Mathias amendment would do just that if it passed.[12]

Under pressure from studios and fearing possible outcomes on both judicial and legislative fronts, video retailers began to band together. Informal local conversations turned into regional groups that met regularly to discuss the industry and strategize together. These meetings were organized on an ad hoc basis—John Pough, a retailer in southern California, formed the Southern California Video Retailers Association (SCVRA) in July 1981 by telephoning ten of his fellow retailers to talk about the rental question. Only six showed up to the first meeting, but within a year there were almost thirty members.[13]

By mid-1981, the particular tension between retailers and studios had reached a breaking point. Warner Home Video had decided to launch its new rental-only plan, pointing to the popularity of rental-oriented video stores as an indication that "the consumer has told us that it's a rental market." According to Warner, going rental-only was merely "an attempt to share in that revenue stream . . . we are looking at this as pragmatic business people."[14] To retailers, though, it looked like an attempt to capitalize on their hard work. One commentator described their reaction as "like colonists who had been through a few tough winters and were about to harvest a bumper crop only to learn that the mother country was going to raise their taxes."[15]

In mid-September, Warner's distribution arm launched the new rental program in Texas. Al Rabe, who owned a store in Plano, was uncomfortable about the weekly cost of leasing individual videotapes as well as the plan's intrusive administrative requirements, but was told that other dealers were signing up quickly, and that the shift to the Warner plan was inevitable. Smelling a rat, Rabe called around to dozens of other area stores, and found that they were actually as dubious as he was. They hastily organized a meeting, and on September 21 more than thirty storeowners met in a Dallas pizza parlor and voted unanimously to oppose the Warner plan. Around the same time, another self-organized group of retailers met in Houston, with all but one coming to the same conclusion.[16] Meanwhile, out in California, George Atkinson called for a "de facto boycott of the 'Warner Rocky Horror Rental Show.'"[17]

Tensions flared even higher at the Las Vegas CES show in January, where storeowners reached a critical mass of anger and frustration with Warner Home Video. In response to the complaints in Texas and elsewhere, Warner revised its rental plan, renaming it "Dealer's Choice" and offering licenses of varying lengths and costs, as well as several tiered categories depending on a Film's popularity. The new plan was introduced at CES, but retailers were less than enthusiastic, with one going so far as to suggest, "Let's pull their booth down."[18] Over the next few months, this frustration and sense of common grievance was channeled into the formation of the Video Software Retailers Association (VSRA), a trade association helmed by Chicago-area storeowner Michael Weiss and including outspoken retailers like John Pough of the SCVRA.[19]

While all of this was going on, distributors like Noel Gimbel were caught in the crossfire between studios and retailers. Through his experiences in the record industry, Gimbel had become active in the National Association of Recording Manufacturers (NARM), and he believed that the video industry needed the heft of a trade organization to improve the retail environment and give retailers a unified voice. A NARM board member, Gimbel saw no essential distinction between video and music, and thought it made sense to let a fledgling video trade association grow under the wing of the larger, more established organization.

Thus NARM hosted its first video-oriented event in August 1981, a three-day convention chaired by Gimbel and attended by more than 400 retailers, distributors, and studio representatives. Conference events included panel discussions of rental versus sale, illegal piracy, and release dates, as well as a speech titled "The Facts of Life in Video Retailing" given by Dr. Theodore Levitt, a professor in the Harvard University Graduate School of Business.[20] Tellingly, the convention opened with a speech by NARM president John Marmaduke titled "What is NARM?" that established a theme for the three days: "NARM is here to put money into your pockets and products into the homes of your customers . . . NARM can only be as good as the information we get from you, the software dealers. We in video have many ideas that can be shared, and while these ideas may be wrong, at least they are always exciting."[21] While the events were no doubt useful to attendees from an informational standpoint, the underlying idea seems to have been to introduce attendees to NARM itself.

Later that year, while the Texans were boycotting Warner Home Video's rental plan, Gimbel flew a handful of video retailers from across the country to Chicago for an informal meeting. The fourteen attendees met in the slick black and silver lobby of Sound/Video Unlimited to discuss the formation of a new organization, and agreed to form a retailers' advocacy group as a division of NARM, with its own policy board and by-laws.[22] The brand-new Video Software Dealers Association (VSDA) was officially founded a few months later, at the same Consumer Electronics Show where some retailers were ready to tear the Warner Home Video booth into shreds.[23]

The VSDA differed from its rival VSRA in several ways—for one thing, the VSDA counted retailers *and* distributors among its members, as opposed to the more militant, retailer-only VSRA; for another, the VSDA was funded by NARM, and had the strength of its lobbying and PR efforts behind it. Their names, however, were confusingly similar, so much so that the VSRA changed its name to the Video Retailers of America (VRA) in April 1982. While there was active competition between the two organizations for the better part of a year, the VSDA quickly emerged as the more powerful organization (thanks in no small part to the institutional clout NARM offered), and by early 1983 many regional associations like John Pough's SCVRA had voted to join the VSDA.[24] Though the VRA limped along and held conventions through much of the 1980s, the VSDA was clearly seen as the voice of the video industry.

Nowhere was the importance of the VSDA more apparent than its annual conventions, the first of which was held in Dallas in August 1982. Picking up where the previous year's NARM convention left off, the Dallas meeting brought studios, distributors, and retailers under one roof to discuss the future of the industry. The VSDA's roots among distributors were apparent, though—as one attendee recalled, "it was actually kind of funny because, even though it was a retail trade association, there were hardly any retailers there".[25] From this first meeting onward, it was clear that important industry business would be foregrounded at the VSDA shows; the big news in Dallas in 1982 was the announcement by Mel Harris, the president of Paramount Home Video, that Paramount was going to offer videotape copies of *Star Trek II: The Wrath of Khan* for a suggested retail price of $39.95, a move loudly cheered by retailers and generally acknowledged as integral to the creation of a consumer videotape sale (rather than rental) market.[26]

Over the next few years, the VSDA show grew, drawing more and more retailers. The shows were held in the summer, since many independent retailers found it easier to take a few days off when schools were out of session and their kids could mind the store for a few days. Booths at the summer VSDA show were cheaper than at the CES show, and the audience was more targeted. Moreover, the "video ghetto" at CES grew less appealing, hitting its nadir at the 1984 summer CES, where video exhibitors were relegated to a "temporary structure,"—a tent on the grounds of Chicago's McCormick Place. According to one report, "It was a tossup as to which was worse: The heat in the unventilated tent or the Lake Michigan breezes let in when the flaps were raised".[27] Following the tent fiasco, many in the video industry left the CES entirely for the VSDA show, only returning in 1986 after much cajoling by the CES.

In a sense, this struggle with the broader consumer electronics world helped create a unified sense of the video industry as a distinct entity. Though relations between many retailers and studios were still strained, in some ways a crisis had been averted through the formation of the VSDA. By bringing suppliers, distributors, and retailers into one big tent, Gimbel and NARM had helped to ease the polarization brought about by the rental plans. Though the VSDA clearly sided with storeowners in those days, studios used it as a space in which to enter into a dialogue with retailers, and vice versa. Moreover, through convention workshops, newsletters, and other materials, the trade organization helped to establish a broad base of common knowledge among retailers across the industry.

Consultants and Franchises

It's important to remember that once the video craze began to catch fire, opening a video store was as exciting as investing in the stock market during the dot-com bubble, and many prospective retailers didn't know the first thing about running a business—some were simply inspired by the sight of customers lined up at the cash registers of their local video store and thought "I could do this!" While distributors and trade associations were a vital information source for early video retailers, most new storeowners didn't tap into those information networks until they had already set up their stores. In order to get from "Hey, I'd like to open a video store" to even knowing how to call a distributor and place an initial order, novice retailers often needed a bit of assistance.

Arguably the easiest way to learn about the video business was to just ask a video retailer how he or she got started. Though some storeowners were protective of their trade secrets, most were gregarious and happy to talk about their industry. When he decided to add prerecorded videotapes to his New Jersey camera stores, Mike Salomon walked right into the main Video Shack store, found a manager, and asked bluntly, "I have a camera store in Hazlet and I would like to add video. Do you have a wholesale distributor or can you tell me who I can get in contact with?"[28] Though storeowners were often receptive to these sorts of queries, sometimes these requests for help were too much. "A problem we've had," reported the owner of Chicago's Video Dynamics, "is that since [our distributors] are very pleased with the layout of our store, when people go to them and say 'We want to open up a video store,' they say 'Go see Video Dynamics first.' Well, we don't want the competition in our store. We don't want to be copied."[29]

While many retailers were willing to give free advice to others just starting in the industry, some storeowners found it could be quite profitable to act as freelance video store consultants, charging for their expertise. Michael Dark, an early video hobbyist who had moved from Las Vegas to northern California and opened a video store, lost count of the number of stores he helped set up between San Francisco and Los Angeles. "We'd have people come [into the store] and say 'I'm going to open a store.' So, for $5000 we'd help you open up a store. We'd go out and do all the initial ordering, start them out with a barcode reader, get them set up with all their cases, blank product, all the stock and accessories. It was a total turnkey operation." In some ways, Dark filled the same role as many distributors, traveling to his clients' stores for the first few weeks they were open and making suggestions on tape orders: "I would get the order sheets and . . . knowing their customer base, how much money they're making in a week, I'd just circle things off and put five VHS here, two Beta there."[30]

Unlike distributors, however, storeowner/video consultants like Dark came from an explicitly retail background, and were particularly sensitive to customer service. "For the first two to three weeks, I would stay in that store and show them, not how to do the business per se, but how to mind their customer base. That was where I really succeeded, teaching people how to listen to what their customer wanted. Ask them, 'How was the selection?' 'What did you like?' Not only does it lets you know if you've got a stinker on your shelf . . . but you know what to order, what the customer wants." This kind of support for new retailers paid off down the road, because "somebody would eventually come to their store and say 'hey I want to open one up,'" leading to even more consulting business.[31]

If the local video store didn't happen to have a Michael Dark on hand, a budding video retailer could opt for an affiliation with a national chain, of which the most prominent was George Atkinson's The Video Station. Atkinson, who had pioneered the idea of videocassette rental when he opened his first store in 1978, began helping others set up stores almost from his first day in business. Part salesman and part entrepreneur, Atkinson was possibly the single individual most responsible for proselytizing the video store boom—in 1983, a fellow retailer said, "[George] introduced the idea of video retailing to hundreds of people . . . he's exposed at least 1,000 serious people to the idea."[32]

One such client of Atkinson's was Ken Dorrance, who was turned on to the idea of video retailing by Mary Ann Black, a Video Station affiliate in Oakland. After showing him around her store, Black put Dorrance in touch with Atkinson, and after a few phone calls and a trip to Los Angeles to speak in person, Dorrance and his wife signed up. For their initial fee, they received a manual explaining how to run a store, a catalog listing available tapes as well as suggestions for initially stocking the store, and the right to use the Video Station name. "Everything else you were on your own to do."[33]

This freedom to run one's business as one saw fit was appealing for many Video Station affiliates. "Because we're independent owners, we set our own policies," said another Video Station affiliate. "I didn't like the idea of a franchise, where someone would tell me how to run my business, but I did like the idea of a nationwide affiliation with an easily recognizable name."[34] In this way, affiliate programs like Video Station's offered what was essentially a commercialized form of the distributor/retailer relationship, giving retailers nonbinding advice on starting and running their stores. In place of weekly sales calls, Atkinson sent out a monthly affiliate newsletter offering continuing guidance and information on the video industry, and the Video Station maintained its own discounted distribution network through which retailers could choose to place their orders. The affiliate model was one of independent, autonomous retailers—as the head of another affiliate program, Steve Garvin of Cleveland's Network Video, wrote, "After franchisees get into the business and you teach them, they don't need you forever. Just because they need guidance once in awhile is not reason enough to charge a royalty."

Ron Berger, however, had a different vision of his role as a video store consultant. In 1980, Berger was operating a chain of franchised camera stores in Portland, Oregon, when a colleague called him up and described the burgeoning video industry. Though initially dubious, Berger was intrigued enough to choose three test cities (Las Vegas, Los Angeles, and Hartford, Connecticut) and investigate their video markets. Concluding that there was "a potentially huge opportunity," Berger hired several people and in August 1980 methodically set about mapping the current state of the industry.[35]

Using reverse directories, Berger determined that there were approximately 900 video stores in existence, none of which were exclusively video (not even Video Station). Berger and his staff called each of these retailers, polling about 200 of them with a series of fifty questions about their stores, and they tabulated the results in a document titled the "First National Video Survey." Then Berger ran a one-inch ad in the *Wall Street Journal* offering copies of this report for $95 each. He sold thousands of copies within a few months.[36]

Immediately, Berger seized the opportunity, drafting franchise offering documents and incorporating National Video in late 1980. He and his wife made a reservation at the Circus Circus hotel, and headed for the 1981 Consumer Electronics Show. Without ever having run a video store himself, Berger ran another ad in the *Wall Street Journal*, this time reading "if you believe that the future of the video industry is wise in franchising, then call this suite at Circus Circus." While some video retailers were agitating against Warner Home Video and Noel Gimbel was creating the VSDA, Ron Berger was giving interviews and passing out media kits in the CES press area. Over the four days of the trade show, his hotel phone rang three times, and all three callers bought franchises. Within a year and a half, there were 323 National Video franchises around the country.[37]

Berger's National Video franchise operated under a far different philosophy from George Atkinson's Video Station affiliations. "What franchising is all about when it's done right," says Berger, "is a really good marketer who has a concept and then finds people who have local retailing smarts, operations ability and capital, and he aligns himself with those people by selling them a franchise. The franchise agreement dictates that they're going to operate their store exactly according to the plan set by the marketing vision guy . . . if he says wear pink uniforms, that's what they're going to wear, and if he says paint your roofs pink, that's what they're going to do. You give [franchisees] training, an operations manual, assistance with site location, assistance with who to hire . . . and of course you give them the basic supply."[38]

In a very real sense, a franchise is a technology that organizes willing store-owners into a homogenous group, codifying the tacit knowledge of retailing into a series of proscriptions and instructions to be followed. Franchisee agreements mandate stability, but they also serve as an enforceable closure mechanism—when a franchisee signs that agreement, he or she is legally bound to adopt the franchisor's frame for the technologies being sold. For Berger, this meant more than simply writing a National Video franchise manual for franchisees—he needed to be able to offer his franchisees information on how to stock their stores that was better than the informal advice distributors offered. His wife Carol was in charge of programming, and she developed a computer-based matrix that rated movies using weighted categories like "star power" and "box office receipts." Over time, this system (dubbed the "Budget Maker") was refined such that a National Video franchisee would simply dial in via modem to a server that offered instructions; "for every thousand dollars in revenue that your store does per month, you need one copy of this, half a copy of this, one-tenth of a copy of this and so on." All that was left to the retailer was to follow the instructions.

Consolidation

Through the late 1970s and into the mid-1980s, the loose norms of individual distributors and video retailers gradually hardened into a defined sense of how a video store should look and what it should offer to consumers. Casual advice from distributors and informal gatherings of retailers coalesced into regional, and then national, trade organizations. Initially, the nascent video industry used the financial and organizational resources of other industries (like NARM and the CES show), but eventually video retailers and distributors created institutions of their own. Meanwhile, entrepreneurs built their own subnetworks through consultancy, affiliations, and franchises (each of which established a more rigidly defined set of norms and practices than the last).

The end result of all this organizational work was a gradual closure for the video industry—in short, the retailers and distributors settled on a common understanding of the business, as well as the nature of the products they offered consumers. As the channels of communication between retailers became more formalized, their sense of what "video" meant became more homogenous. In a sense, discussions about what video was and what a video store should be moved from the level of the interpersonal (individual conversations) to the mass (via trade organizations and franchises), as storeowners from myriad backgrounds settled on a common identity as video retailers.

As a disciplinary technology, franchising can only dominate a market when it exists within a relatively homogenous culture. The centralized decision-making and uniformity that are core components of a franchise assure consistency across various locations, but they also reduce the ability to adjust to local norms and expectations. Thus a franchise only succeeds when it can leverage a preexisting frame that defines a consumption junction across geographic spaces. The hamburger fast food restaurant, for example, thrived as a franchise in the automobile-centric world of southern California early in the twentieth century, but wasn't successful on a national level until that car culture had spread across the country in the 1950s.[39]

By the mid-1980s, the video industry existed as a diverse ecology of national chains like Video Station and National Video, smaller regional chains like Washington DC's Erols and the Midwest's Movies-To-Go, and myriad independent mom and pop stores that had sprouted everywhere. Everybody knew his or her place in the broader industry, and their investment in and dependence on the network of franchises, organizations, and other ties began to accumulate what Thomas Hughes calls "technological momentum," solidifying and reinforcing the status quo.[40] Within the next few years, however, the industry would be rocked by one company that used the technology of the franchise to radically reshape the landscape of home video: Blockbuster Video.

Blockbuster

Florida businessman H. Wayne Huizenga, who had made his fortune in waste management, "rarely watched movies and didn't own a VCR."[41] He associated video with the "sleazy stores in bad neighborhoods" that had been a hallmark of the earliest purveyors of pornography and bootlegged tapes, but in early 1987 was convinced by a friend to drop by a store belonging to the then-fledgling Blockbuster Video franchise. Huizenga was surprised by the store design, which was clean and bright, as well as by its particularly extensive selection of tapes (the first Blockbuster Video store had opened in 1985 with over 8,000 titles on its shelves). Sensing an opportunity, Huizenga became an investor, and by April of that year had taken over as chairman.

The Blockbuster strategy was simple—pump as much money as possible into buying local and regional chains while keeping centralized control over the look and feel of individual stores. By the VSDA convention the following year, Blockbuster had acquired two other chains and its more than 250 stores dotted the country. At the convention, Huizenga's marketing executive Tom Gruber outlined a vision for the future of the company, and it was expansive. Gruber had spent eighteen years working for McDonald's before joining Blockbuster, and both he and Huizenga were explicit: Blockbuster wanted to be the McDonald's of home video (the comparison was so deliberate that at one trade show presentation, huge photographs of Huizenga and McDonald's leader Ray Kroc were projected side-by-side).

Essentially, Blockbuster Video played off the technological frame for home video that had been established over the past ten years, structuring its bright, well-packaged stores around a simple idea: mainstream Hollywood movies. No hardware, no accessories, no photo processing or other side businesses, and no food except for candy, soda, and the ubiquitous popcorn.[42] Perhaps the clearest example of Blockbuster's positioning with regard to the rest of the industry was its slogan, "Wow, What a Difference!," a deliberately cheery jab at the image of the independent video store with its limited selection and morally ambiguous back room. The franchised Blockbuster stores even guided employee-customer relations by establishing a greeter in the front of the store whose job it was to say "Hello, how are you? Welcome to Blockbuster," and the franchise manual included instructions on how to say goodbye to every customer as they left. At the same time, however, employees were instructed to refer customers to printed recommendation materials (provided by the franchise) rather than offering suggestions themselves.[43]

In an odd twist of fate, Rich Nathanson, who had worked for New Video for years, wound up as an assistant manager of the first Blockbuster store in New York City and was able to make a firsthand comparison of the new franchise with earlier video stores. At the time, Blockbuster had honed the store construction process, with everything necessary to build and stock a store packed in a tractor-trailer in the order it would be needed, and the transition from empty storefront to ready-to-open store could take just a single day. As much as Nathanson had loved the employee and store culture at New Video, he was stunned by Blockbuster's efficiency and forward drive. "It was an amazing company," he explains. "I'm not saying good or bad, but I'm just saying amazing."[44]

By the early 1990s, Blockbuster Video was the dominant player in video rental, with several national franchises following in its footsteps. Mark Wattles, the founder of Hollywood Video, remembers seeing a Blockbuster store for the first time and thinking, "that store [is] going to wipe out the industry as we knew it."[45] Thinking along the lines of fast food, Wattles knew that most heavily franchised markets tended to have more than one player, so he decided that his Hollywood Video would be the Burger King to Blockbuster's McDonald's. Franchisers like Wattles and Huizenga were businesspeople who saw an opportunity, and no love of movies inspired them— their interest lay fundamentally in the functioning of the corporation and the market. Ultimately, the men who ran the two dominant video store empires were more enthralled with the technology of the franchise than the technology of video.

As for the independent retailers, who had in many ways invented the norms on which Blockbuster capitalized, they mostly faded into the background. Some sold their stores to the larger franchises; many more simply went out of business. However, in some parts of the country, particularly the very urban and the very rural, nonfranchised stores carved out their survival.[46] Blockbuster and other chains dominated the suburbs, but found it much harder to make inroads into major cities, and not profitable enough to open stores in the country. The family-friendly no-pornography policy of Blockbuster, Hollywood, and other chains also helped independents survive, as customers had to turn away from the mainstream franchises in order to rent or buy adult videos. Meanwhile, in the centers of major U.S. cities, a handful of specialty video stores survived as artifacts of an earlier era. Video Droid north of San Francisco, The Video Room in New York, TLA Video in Philadelphia, Chicago's Facets, and stores like them were lasting reminders of a time when the atmosphere in a video store was communal rather than transactional, and a customer wouldn't get an odd look for asking an employee "What do you think I should watch tonight?"

7 The Thin Line Between Movie and Technology

This book has mainly focused on the story of video stores, but what of the movies themselves? One might read the narrative I've laid out and assume that movies as a form remained relatively unchanged by entering this new retail space and new medium, but this was not the case. While Fred Wasser has persuasively argued that the rise of video directly impacted the *kinds* of movies that get made, I want to briefly shift gears, into a more media studies mode of analysis, examining what happened to the actual form of movies as texts once translated onto video.[1]

In its first few decades, the U.S. motion picture industry had incorporated specific technologies and social practices into a system stretching from cameras to theater projectors, from Hollywood studios to audiences around the world. As the public face of this sociotechnical system, the movie theater was the institution that governed most filmmakers' and audience members' experiences of movies, and when alternate ways of viewing were introduced, they were framed as "ancillary media," understood in terms of how they differed from the primary medium, the movie theater.[2] By 1980, viewers could choose from several ways of watching a given movie, each with its own distribution network that mediated the movie for the viewer.[3]

Though the practices of actors, directors, and film crews were essentially the same across these three media, once their work was done the network of social relationships and technical artifacts that constituted "moviemaking" trifurcated, branching off into the three disparate directions of television, video, and the theater.

In *Understanding Media*, Marshall McLuhan wrote that "the content of any medium is another medium," describing a teleological relationship between a given medium and the media that came before it, precisely the sort of relationship that seemed to exist between the theater and video.[4] Following this logic, a movie-on-videocassette simply "contained" a movie-in-the-theater, and thus was understood in terms of the theater rather than on its own terms. Like much of McLuhan's work, however, this aphorism is founded on the inherent properties of artifacts, and is thus irreconcilable with the view that meaning around technologies is socially constructed rather than innate.

While one (if feeling charitable) *could* read McLuhan as pointing to our tendency to understand technologies in terms of their historical predecessors (what he himself referred to as the "rearview mirror"), theorists Jay David Bolter and Richard Grusin offer an alternate approach, validating the idea that media "remediate" previous media—while at the same time leaving behind the implied technological determinism in the theory—by conceptualizing media as hybrids of social, economic, and technical components.[5] In their book *Remediation*, Bolter and Grusin chart the various strategies by which digital media remediate their predecessors, most notably by transparently offering access to the content of the predigital media ("immediacy") or by explicitly (and often ironically) foregrounding the differences between the new medium and the old ("hypermediacy").[6]

Bolter and Grusin's elaboration on these rhetorical stances is useful, but it never seems to give up on that teleological explanation of remediation—for them, there is always a new medium that is remediating the old. By taking a theoretical view of medium and message as coproduced rather than as distinct ontological categories, I want to offer a more symmetrical view of the relationship between media, situating my analysis from the user's perspective (for whom no privileging of one medium over another is inherent to the technology itself). From this perspective, sitting in a theater and watching television are two distinct technological practices, and the movie that one watches is simply a construction, coproduced along with the inert medium during the practice of "watching the movie" in either context. If there is no message other than that coproduced through media use, then the notion of movies on television or video as ancillary media isn't an inherent property of the media ecosystem, but rather an added meaning, layered on top of the various technologies in question.

This perspective raises an intriguing problem—if the movie as a text is coproduced through the actual act of using a media technology, how does it maintain a coherent identity as it moves from one technology to another? Viewers experience each medium as having a distinct set of social and technical practices, each laden with its own cultural meanings and norms. The experience of going out to a theater and watching in the dark with hundreds of strangers was quite different from that of sitting in one's well-lit living room, possibly with a partner or child; much has been made, for example, of the fragmented quality of home viewing thanks to the myriad interruptions of domestic life.[7] The movie (the "message") is encoded differently in each system, using chemicals on celluloid, electrons on magnetic tape, or electro-magnetic waves broadcast through the air. Even neglecting questions of the relative size of the screens, theater and television images were produced in radically different ways, and both psychologists and cultural theorists have speculated on the cognitive difference between projected light bounced off a reflective screen and radiant light from a cathode ray tube.[8] The actual images weren't even the same: one second of theatrical film involved twenty four consecutive images, while the same second on a television screen required six more. From the perspective of the sociotechnical system as experienced by the user, there was strikingly little in common between the practices of watching a movie in a theater, on TV, or on video.

Users, however, faced with their experiences of these radically different sociotechnical networks, found the consistency of movies across media remarkably unproblematic: *Jaws* on television was the same text as *Jaws* in a theater, or on video. If not from within the medium itself, from where did this consistency come? In each case, viewers constructed a version of *Jaws* through the practice of viewing, but this movie-in-practice was informed by an idealized expectation of what the movie *should* be like, an expectation that led users to mark the boundary between medium and message at a precise point. This socially constructed archetype (which I'll call The Movie) is rather like a Platonic form, an idealized text existing only in the abstract, but appearing in various imperfect physical manifestations.[9]

This situation wasn't unique to movies-on-video, and is in fact latent in any media technology. We can only separate radio static or newspaper typos from the message itself if we have a preexisting concept of what that message *should* be like—a idealized version of the message that we compare to the manifestation at hand to determine where to draw the line between medium and message. Our understanding of the message is, in short, an essential component of the technological frame through which we view the medium. While this process is invisible in the context of stable media technologies, the construction of movies-on-video relative to the theater or to television broadcasts required an active negotiation both in video stores and in the broader cultural discourse, not just of the technologies' meaning or use, but also of the very nature of the Movies mediated by those systems.

The Trappings of Television: Commercials and Editing

Movies themselves enjoyed a life on the television screen well before the debut of home video technology. Hollywood motion pictures were an important part of many 1950s programming rosters, and were often the only programming option available to independent television stations trying to compete with the major networks.[10] As television became entrenched in American life, the afternoon matinee, prime-time feature, and late-night movie became standard parts of the television schedule. In 1976 alone, according to *TV Guide*, "the networks announced the acquisition of 72 new theatrical films and the scheduling of four movie nights a week. By midsummer they had offered more than 140 premieres and the scheduling of four movie slots spread over all seven nights of the week."[11]

While a vital part of the broadcasting day, movies on television were always just that: movies *on television*. In television listings and other discourse, motion pictures were often preceded with the modifier "theatrical," to distinguish them from their cousins, the "telefilms," which were quite literally "made-for-TV." A movie on television was understood as a qualitatively different product than a movie in a theater, with the former a mediated and altered version of the latter. In 1982, for example, *New York Times* film critic Janet Maslin wrote:

A censor made mincemeat out of "Dirty Harry" several Saturdays ago, when an outrageously overedited print was shown as a late-night television movie. This wasn't simply a case of snipping out the odd bit of colorful language here and there, or of protecting the home audience from excessive gore.

Scenes had been trimmed wholesale, whether they required trimming or not. For instance, the contents of a ransom note had been bleeped out, thus making the whole plot close to unintelligible.

Harry's famous speech beginning "I know what you're thinking—did he fire five shots or did he fire six?" had disappeared, and with it the key point that this detective always knows how many bullets are left in his gun. The action scenes had been doctored so artlessly that the mayhem seemed entirely random, and thus even more violent than it had been in the first place. From beginning to end, the censor did an inexcusably senseless and slovenly job . . .

Of course, "Dirty Harry" isn't alone in being treated this way. As any moviegoer with a memory knows, the dread words "Edited for Television" mean that a favorite scene in a favorite film may or may not turn up in the television version. If it does, it may be unrecognizable. Leading actors may sound as if they've developed hiccups, once bits and pieces of their best-known speeches have been chopped out and the leftover words carelessly pasted together. Television censors lop and slice with impunity—sometimes to make time for commercials and sometimes to sanitize—never having to worry about what the traffic will bear. Until very recently, the traffic—the at-home film fan—hasn't had much choice.[12]

Maslin's condemnation speaks to the peculiar status of motion pictures on broadcast television at the beginning of the video store era. The most perfect version of The Movie was generally acknowledged as that which one could experience in a theater, the original and intended home of The Movie, as opposed to the ancillary television market, where a movie might be "edited for television" or interrupted by commercials.[13] When a television station *did* broadcast a movie without edits, it was usually a distinctive (and sponsored) event, like the following: "At KCET-TV in Los Angeles, Xerox is sponsoring a 26-week 'Film Odyssey,' bringing to public television audiences (and this will be nationwide) feature films from all over the world, shown uncut and unmangled by commercial interruptions."[14] While "cut and mangled" movie broadcasts on television often received high ratings, this construction of the idealized motion picture cast a shadow over the experience of watching for many viewers, who—like Maslin—were quite aware that they weren't getting the real Movie. However, the constraints of a given theater (small screen, bad projectionist, noisy neighbors) meant that even the movie-in-the-theater might not be a complete representation of The Movie itself, though invariably a more accurate one than the movie-on-television.

As we've seen, video stores explicitly framed movies on videocassette in terms of the theater rather than the television set. It was crucial for video producers and retailers to distinguish their products from the preexisting meanings of movies on broadcast television, to make the movie-on-video preferable to the movie-on-television (or, to shoot the moon, preferable to the movie-in-the-theater itself). Even though some of the same technical transformations were bound up with creating movies-on-video and movies-on-television, they were reinterpreted and reconstructed in different ways.

One of the adjectives most often applied to prerecorded movies on videocassette was "uninterrupted," a coded way of saying "without commercials."

Since movies in the theater didn't stop for commercials every fifteen minutes, the commercials that were all too common on network movie broadcasts were understood as interruptions imposed on The Movie, artifacts of the particular system of advertising-supported broadcasting dominant in the United States since the 1920s.[15] The efforts of the videophiles described in chapter 1 point to this tension; their practice of editing commercials from broadcast movies was a clear attempt to create a more faithful rendition of The Movie from the imperfect, commercial-laden broadcast. A prerecorded cassette without commercials was thus closer to the ideal of The Movie than one with commercials. "The reason I buy [prerecorded] videocassettes is that I don't have to deal with interruptions by commercials," said one VCR owner in a 1985 survey. "That's why you buy the tape," said another, "'cause it's like you're at the movies."[16]

While this lack of commercials was one of the biggest selling points of prerecorded movies on videocassette from the earliest days of Magnetic Video onward, the VCR was understood as a commercial-free technology in another, subtler way. One of the most commonly mentioned uses of the VCR was to fast-forward through commercials when watching time-shifted broadcast programming.[17] Though such "zapping" caused a panic in the advertising industry, this freedom from commercials was a core part of the VCR's identity, which bled from its time-shifting uses into its characterization as a movie playback device. Commercials seemed fundamentally against the ethic of VCR use, and one retailer's statement that "I am against commercials on videotapes as it bastardizes the videotapes . . . people watch videotapes to avoid commercials" could have just as easily been understood as a reflection on time shifting as its original context.[18]

On the other hand, commercials had proved a lucrative source of revenue for the broadcast industry, and Hollywood studios found their allure difficult to resist. Early videocassettes included trailers for other videos, essentially commercials, but these weren't seen as intrusive since trailers had been an established part of theatrical movie presentations for decades. By the mid-1980s, however, the Hollywood majors were beginning to murmur about the possible benefits commercials on videotapes could bring to consumers, most explicitly lowering the price of movies on video to "as low as $10 or $15."[19] The taboo was finally broken with the 1986 video release of *Top Gun*, which included a Pepsi commercial before the feature presentation. The cassette's high sales figures indicated that consumers were willing to tolerate the commercial intrusion in return for a dramatically cheaper cassette ($26.95 in the case of *Top Gun*), so long as the commercials didn't violate the integrity of The Movie itself. They were seen as a necessary evil, an artifact of the sociotechnical system of video production in the same way as television commercials were an artifact of the broadcast networks.

Over the next few years, commercials became a stronger presence on prerecorded videocassettes, though always before or after the movie, never crossing that invisible boundary between medium and message. Their advertising message began to leak from the tapes into the video stores in which they were sold, blurring the line between the movie and the product being advertised. For example, Frederick Wasser writes that when Vestron released *Dirty Dancing* on video with an ad for Nestlé's white chocolate preceding the movie, "Video retailers were urged to stock Nestlé candy bars and Nestlé promoted *Dirty Dancing* in its own advertising."[20] The particular manifestation of *Dirty Dancing* on video was embedded in a larger nexus of consumer products, both in the consumption junction and in the larger culture. Ultimately, however, storeowners did find ways of resisting the encroachment of consumer culture into their stores, or at the very least actively reconstructing it in a preferred form; one retailer vowed not to let Pepsi's initial ad go unanswered. "We're going to take the commercial off [of *Top Gun*] and put our own commercial over it. I'm not getting any residual for selling Pepsi. I'd like to tell Pepsi to go and buy a Coke."[21]

Pan and Scan versus Widescreen

Perhaps the most obvious transformation that a movie undergoes in the transition from film to video is its physical shape. In the 1970s, theater screens and television sets were simply shaped differently; the aspect ratio (the proportion of the width to the height) of a standard television screen was 1.33 (4:3), while that of the average theatrical film might vary from 1.85 to 2.35. This disparity meant that the film image had to be quite literally remediated to fit on the narrower television screen.

For the first part of the twentieth century, motion pictures were shot and projected in a 1.33 aspect ratio that had grown out of the technical and aesthetic norms of the industry-standard 35mm Eastman film.[22] By the 1930s this 1.33 aspect ratio was so entrenched that when the addition of a soundtrack along one edge of the film strip necessitated the narrowing of the image, many projectionists simply masked the top and bottom of the projected image to maintain their preferred shape for a movie. As film historian John Belton writes, "this practice alarmed Hollywood producers, who found that the heads and feet of actors and other important elements of the picture were being cut off in projection," and the Academy of Motion Picture Arts and Sciences voted in 1932 to restore the traditional 1.33 aspect ratio by shooting with "specially-cut aperture plates in both cameras and projectors," establishing this so-called "Academy format" as the official industry standard.[23]

While television's early history was marked by myriad controversies over the frame rate or number of vertical lines that should constitute the broadcast standard, there seems to have been no real dispute over the appropriate shape of the screen. The Academy format was unproblematically imported into the context of home television technology, and when the National Television System Committee (NTSC) made a formal standards recommendation to the Federal Communications Commission, the proposed ratio of the width to height was 1.33. Television and film shared more than simply the shape proscribed by their respective standards—many television programs were filmed for later playback or archiving using a variant of the motion picture camera called the Kinescope, a process facilitated by the identical shape of the televised and filmed images.

Through the 1950s, as Americans began to spend increasing amounts of time watching television in their homes, movie producers and exhibitors found their revenues dropping. In an attempt to distinguish itself from television and other domestic forms of entertainment, the theatrical movie industry constructed a new identity for itself based in large part on the spectacle of the theatrical movie-going experience. While some producers experimented with 3-D and gimmicks like "Smellovision,"[24] this reconstruction of the properties of movies-in-the-theater centered on the screen, in particular its size and shape relative to television (and earlier theatrical) screens. New standards such as Cinerama and Cinemascope offered audiences filmed images that filled their field of view in previously unheard-of ways, with aspect ratios more than twice as wide as the Academy format, and filmmakers made use of the new technology in a wave of epic films including *The Robe*, *Lawrence of Arabia*, and *Ben Hur*.

Thus, by the 1970s the position of the theater as the true and ideal home of spectacular motion pictures was unquestioned. That's not to say that movies were never broadcast on television; they most definitely were. Before being broadcast, however, the actual shape of newer movies had to be transformed from the wider aspect ratio of the theater to the narrow 1.33 aspect ratio of the television screen. In the prevideo days, more or less the only option was to simply center the television camera on the projected film image, chopping off its sides and hoping for the best. While such cropping was necessary, the process often produced rather jarring results (particularly in the case of movies that deliberately made use of the entire Cinemascope frame).

The adoption of video technology by broadcast television stations, however, allowed for a more subtle approach. Technicians created 3/4" videotape versions of films using a machine called a Telecine, essentially a modified video camera that converted a filmed image into electrical signals. In addition to tweaking the color and timing of captured video, Telecine operators were able to pan across the widescreen image so that the final product would (theoretically) include the most important elements of a given shot, wherever they might be on the original. "We technicians would pan back and forth as gracefully and unobtrusively as possible," recalls one Telecine operator, "to show whoever was talking, to tell the story, to present a picture that was composed reasonably well, artistically."[25] In a sense, the Telecine operator was an uncredited author, as responsible for the final image of a movie-on-video as the cinematographer or director.

Such "pan and scan" video transfers were the default for early consumer prerecorded videocassettes, and this particular method of altering the theatrical image naturally went unmentioned in early discourse about movies on video. Recall the original Magnetic Video "Video Club of America" advertisement from chapter 2, and its claim to potential customers that "You now have the opportunity to select and show the original full-length uncut versions of Hollywood's finest movies on your own TV set."[26] The new rhetoric of movies-on-video was intended to foreground their advantages over movies-on-television, and since both forms were by default panned and scanned, that particular method of altering The Movie remained unmentioned in video stores.

Unmentioned, that is, until the mid-1980s. In 1983, Telecine operators at Modern Video Film (one of the pioneers in film-to-video transfers) created a pan-and-scanned video transfer of Woody Allen's *Manhattan*, which was originally filmed in the 2.35 Cinemascope aspect ratio. At the time, virtually all transfers were performed without the permission of directors, many of whom seem to have seen ancillary media such as television or video as irrelevant to their artistic expression.[27] A clause in Allen's contract, however, required that he personally approve any and all versions of *Manhattan*, so Modern Video Film sent him a copy of the cassette. A week later, the company received a faxed message: "Mr. Allen finds this transfer unacceptable. Do it again." Over the next weeks, Modern Video Film created at least three separate transfers, each time sending the product to Woody Allen and receiving the same response.[28] Finally, one technician declared "Maybe he just doesn't like pan and scan" and, with the permission of the now-desperate studio executives, created a transfer that included the *entire* Cinemascope image, with empty space at the top and bottom to simulate the more oblong shape of the theater screen. As one of the Telecine team remembers, "A week of silence, a fax came back, and it simply said, 'Mr. Allen approves this transfer.' And that was it."[29] Later that year, *Manhattan* shipped on videocassette and RCA videodisc with gray bars on the top and bottom of the screen.[30]

This new style of video transfer was dubbed "letterboxing," because the result is rather like looking out through a mail slot. (Telecine operators, followed by the industry more generally, adopted the British "letterbox" rather than the American "mail slot" because the most popular film transfer devices were manufactured by the British firm Ranks and Tell, who had labeled that particular button with the Anglicism). Since pan and scan had been established as the norm for movies on video, letterboxing (though simply an alternate way of transferring from film to tape) was usually cast as an "enhancement," and through the late 1980s it was identified with higher-end video, particularly Pioneer's prestigious Laserdisc format, which had staked an identity as the format of choice for enthusiasts who were creating high-end home theaters. Some stores used the cachet of letterboxing to reinforce their identity as a place for movie lovers; one employee at San Antonio's Bjorn's Disc Store recalls that his store "was the one place where a videophile could get films in letterbox . . . People used to come from all over Texas to buy and rent from us—I had customers from as far away as Norway whom I did business with over the phone."[31]

As letterboxing became more common, however, consumer reactions to the black bars on the top and bottom of their television screens threw into sharp relief their preexisting understanding of the nature of televised images. A televised image had historically filled the entire 4:3 screen, with a literal border between medium and message inscribed along the edge of that screen.

Now, however, only a portion of the screen was intended to be seen as message, and the black bars at the top and bottom were to be understood as an artifact not of the movie, but of the television set—no more meaningful than the wood cabinet in which the screen was housed. Many users who had internalized an understanding of a television picture's natural shape as filling the entire screen interpreted these black bars as somehow blocking or detracting from their ideal image.

On a fundamental level, the way that a viewer perceived the black bars of a letterboxed movie pointed to the technological frame through which he or she understood the practice of watching a video; if they first and foremost understood the practice of using a VCR and prerecorded video as *watching a movie*, the bars made perfect sense, since they helped to bring the viewing experience closer to the ideal movie-watching experience of the theater. On the other hand, other viewers found the bars irritating or frustrating because they understood the practice of watching a video as *watching television*, the norms of which dictated that the image must fill the screen, and in fact that movies might be modified to fit the constraints of the television set.

One of the major spaces in which this controversy played out was the video store itself, and storeowners and clerks often found themselves having to educate their customers on this new way of using their television sets. "If I had a dime for every time that a customer came in and complained that there was something wrong with their tape because there were black bars on the top and bottom," recalls one video clerk, "I wouldn't have had to work at the store at all."[32] Those behind the counter, who tended to strongly understand the act of watching a video as *watching a movie*, tried to explain to customers that letterboxing in fact made the televised image *more* movie-like, but were often unsuccessful. It would not be until the advent of DVD in the late 1990s that letterboxing would achieve a more mainstream identity as a technological practice.[33]

Colorization

The previous sections described various ways in which the technical transformations involved with the move from film to video constructed this remediation as transparent, framing a movie on videocassette as a representation of The Movie that was just as faithful as that same movie in a theater. When several corporations began experimenting with colorization, however, the controversy shifted from whether movies on video were as good as movies in a theater to whether they could actually constitute an improvement.

In 1985, Hal Roach Studios announced that it would begin using a new process, developed at its Toronto-based subsidiary Colorization Inc., to release colorized versions of black-and-white movies from its library of more than 1000 classic films (which included many Laurel and Hardy comedies as well as John Wayne westerns and Frank Capra films).[34] In principle, the process was nothing new; since the early days of Edison, filmmakers had experimented with the addition of color to black-and-white prints, employing teams of artists to hand-color the film, frame by frame.[35] The new colorization process, however, promised a revolution not of kind, but of scale, using computers and video equipment to automate the tedious and labor-intensive process. At Colorization Inc., teams of art directors researched the historical context and mood of black-and-white films to determine the appropriate color palette for a given scene. Once decided on, those colors were given to a team of technicians who used computers to map individual shades of gray from the video masters to the chosen colors, resulting in a colorized version of the original film. As a Hal Roach spokeswoman put it, "It's like a child's coloring book,"[36] albeit one that cost a quarter of a million dollars (per film).[37] Technically speaking, the computerized colorization process was enabled by video technology—the instantaneous and mechanized substitution of shades of color for shades of gray was only possible if the movie in question was encoded as relatively malleable electrons on magnetic tape instead of the fixed chemical grain on celluloid.

Perhaps the greatest (and loudest) proponent of colorization was the broadcasting mogul Ted Turner, who in 1986 purchased the libraries of MGM, RKO, and early Warner Brothers, in large part to ensure programming for his new cable ventures. Shortly after acquiring the libraries, Turner released a list of more than 100 films he had slated for colorization, including classics such as *Casablanca, The Maltese Falcon, Yankee Doodle Dandy, A Night at the Opera*, and *The Postman Always Rings Twice*.[38] Turner framed his decision in the simple economics of television advertising, where color broadcasts claimed higher advertising rates, and taking a firm stand he declared, "If they're going to be on television, they're going to be in color . . . All I'm trying to do is protect my investment."[39] Others in the industry agreed with this economic rationale. "People who buy movies for distribution and sale—television stations, networks, cable television, and so on—always classify the black-and-white movie as a lesser picture, and therefore don't pay as much as they would for a color picture," explained the chairman of Hal Roach Studios. "So we thought, well, if these pictures were in color, they'd command a much bigger price."[40]

Consumers had been encouraged by decades of advertising discourse about the glories of color television to prefer modern, high-tech color images to old-fashioned, quaint black and white, and as we've seen, this discourse was picked up by prerecorded videocassette distributors once mainstream Hollywood films began to supplant the early market in public-domain (invariably black-and-white) movies.[41] While useful for constructing an image of one's technology or content as contemporary, and thus more desirable, such rhetoric had the unfortunate effect of dampening interest in the studios' greatest asset, their collective library of approximately 17,000 black-and-white films and 1,400 black-and-white television series.[42] As many in the industry saw it, colorization promised a larger audience for these movies and television shows without the effort of reeducating consumers to appreciate black and white.[43] Since almost no syndication market existed for black-and-white films, a Hal Roach vice president argued, "We're giving the viewers the option of discovering these wonderful classics that otherwise they would not get the chance to see and enjoy and experience."[44]

According to some filmmakers, however, television and video viewers were *not* in fact getting the chance to see, enjoy, and experience a movie once it was colorized, and they mobilized a massive public relations campaign to protest what one called "an ugly practice, totally venal, anti-artistic and against the integrity of every film maker."[45] Colorization, their argument went, was an extraneous technical manipulation, an imposition on the artistic vision of any filmmaker who had created a movie in black and white. For Woody Allen, "it [was] the same as actually re-cutting a film,"[46] and Ginger Rogers described her reaction to a colorized version of *42nd Street* by saying, "I can tell you how it feels, as an actor, to see yourself painted up like a birthday cake on the television screen. It feels terrible."[47] Rogers's comment, along with many others, characterized colorization as a process that "painted" The Movie, desecrating it by presenting a fundamentally altered text to viewers, with the implication that a colorized version was less true to The Movie than the original black and white.

Rather than engage in a dispute over the relative fidelity of a colorized movie-on-video, colorization proponents offered the radical counterargument that colorization foes misunderstood the nature of The Movie itself. As an editorialist in a trade publication wrote, "It's true that some films were purposely shot in black and white, but it's also obvious that many of the old black and white films weren't produced in this style for artistic purposes, but because of the lack of color technology or simply the prohibitive cost of filming in color."[48] Following this line of argument, the celluloid version of *42nd Street* that audiences had been watching in theaters for decades was an unfaithful manifestation of that Movie, an artifact of an imperfect technical system only able to reproduce black and white. This was the historic preservation argument turned on its head—as a letter to the editor in the *Los Angeles Herald-Examiner* put it, "If [colorization opponents are] right, then old, silent movies should be shown using a hand-cranked projector."[49]

A 1987 advertisement for CBS/Fox Video made the case for colorization even more explicitly, offering still images from four popular black-and-white movies, *Yankee Doodle Dandy*, *42nd Street*, *The Maltese Falcon*, and *Captain Blood*. A square is inscribed through the middle of these stills, within which the black-and-white images have been colorized. A caption below the images proclaims, "Revealed in their true colors!" While Ginger Rogers would have claimed that the area within that square contains color layered on top of the Movies, the caption indicates that it marks the space within which the black and white has been peeled away, leaving the "true colors" of The Movie for viewers to enjoy.[50]

As theorists, we can understand the controversy over colorization as a dispute between two different constructions of The Movie. Consider the case of *Topper*, the first film colorized by Hal Roach Studios; the colorized version of the movie was either an improvement or a defacement depending on the frame through which one understood *Topper* as a Movie. Cary Grant, the star of *Topper*, said, "It was a curious experience to watch a film which I knew to have been shot, and distributed, in black and white, yet was being shown in a colored version that seemed, as far as I can remember, faithful to the original colors of the sets, costumes and countryside."[51] Describing a scene in the same colorized version of *Topper*, however, film critic Vincent Canby wrote, "In Norman Z. McLeod's black-and-white film, Marion's dress and shoes appear to be silvery white. They're the sort of expensive, impractical clothes that, having been worn once, must be packed off to a cleaner . . . Be prepared, then, for the so-called "colorized" version of *Topper* . . . in which Marion wears an evening dress and shoes tinted a shade that might best be described as grayish lingerie-pink."[52]

Figure 7.1
Advertisement for CBS/FOX colorization. (The portion of the stills inscribed within the white box was rendered in color, while the rest of the ad contained only shades of gray.)

Canby and Grant reacted to the colorized version of *Topper* differently because each compared that specific manifestation of the movie to a different idealization of what it *should* look like. Canby, whose only experience of *Topper* was as a black-and-white film, saw color as detracting from the movie's meaning, a "ghastly process" that "distorts history." Grant, on the other hand, compared the images on the screen to his own experiences on the movie set, and reached the conclusion that the colorized version was in fact a more faithful representation of, say, the shoes that Canby preferred in black and white. For Grant, the colorized *Topper* was a more faithful reproduction of actors performing on a set, while for Canby the black-and-white version was a more faithful representation of what he perceived as the director's artistic intent.

Another critic, Andrew Sarris of *The Village Voice*, went a step further than Canby. Describing his frustration when watching colorized films, Sarris wrote, "Truth to tell, my research has been hampered by my inability to sit through a colorized movie for its full running time. I have tried with *Topper, The Maltese Falcon, It's a Wonderful Life, Miracle on 34th Street,* and *Yankee Doodle Dandy*, but sooner rather than later, I turned off the color on my set to get a grayish version of the original print."[53] In turning down the color (a common suggestion offered by colorization proponents), Sarris actively reshaped the movie on the screen to more closely match his idealized conception of the black-and-white films in question.

While colorization was fundamentally bound up with video technology, such skirmishes over colorization initially centered on movies-on-television. Ted Turner's widely advertised colorized broadcasts of *Yankee Doodle Dandy* and *A Miracle on 34th Street* were the first time that most Americans saw such a colorized image. Additionally, broadcast television was the market offering the highest incentives for companies to colorize specific films, since potential syndication fees were far greater than expected revenues for prerecorded videocassettes. Colorization made the front page of newspapers like the *New York Times* and the *Los Angeles Times*, and Ted Koppel devoted an episode of his evening news program *Nightline* to a discussion between director Martin Scorsese, critic Gene Siskel, and Roger Mayer, then-president of Turner Entertainment Company, on colorization's merits.[54] Because the first colorized films debuted on television, the bulk of colorization's framing and definition took place on the mass media level, and customers had already formed an understanding of their place in the media ecology by the time colorized tapes really began to appear in video stores.

That's not to say that video storeowners had no power to influence the construction of colorization; they did, but through 1986 and 1987 storeowners seemed fundamentally ambivalent on the subject. Those with whom I've spoken don't seem to recall colorization as a big deal, and the trade publications of the time mainly echo the corporation versus filmmakers framing of the issue. For storeowners, who were in the odd position of having sympathies for both the business instincts of Ted Turner and the artistic claims of filmmakers like Woody Allen, the potential rentals and sales that colorization might bring were offset by their appreciation and enthusiasm for film as an art form. "This industry was founded on the silver screen," said Key Video's Herb Fischer. "Personally, I'd want it both ways, but if colorization succeeded, it might be pretty hard to find those movies in black and white . . . Me, I'm fighting my emotions."[55]

Movies across Media

Whether cropped, cut, colorized, or "commercialed," movies on video called into question the traditional conception of movies in the theater. Before the VCR, movies seen outside the theater, whether in living rooms or on airplanes, were unproblematic; they were altered from their true form (that which is seen in the theater) in order to fit into those other technological systems. Once movies on video were constructed by video distributors and storeowners as tantamount to movies in the theater, viewers were faced with a dilemma. If a movie on video simply looks different from a movie in the theater but is not framed as inferior, then which is the real movie and which is the altered version?

As I have argued, from a removed perspective there is no one real movie, because the movie cannot exist independent of the technological frame of its medium. Judgments of "realness" or the relative prestige of movies in different media aren't just judgments of a filmic text, they are assertions about the meaning of a specific technology, as well as its relationship to other technologies intended to achieve roughly the same purpose (that is, to show a movie to a viewer). Thus the disputes over cropping, colorization, and other technology-driven changes can be better understood as negotiations regarding a specific technological controversy, that between the VCR and the theater. As the relationship between the two parallel media shifts, so does our understanding of the movies that they brought us.

Epilogue

Now you can watch anything you want to watch anytime you want to watch it. Because Sony's revolutionary Betamax deck—which hooks up to any television set— can automatically videotape your favorite show (even when you're not home) for you to play back anytime you want . . .
What power!
—Sony advertisement, 1975[1]

"You cannot program your VCR without a PhD in electronics."
—VCR owner, 1988[2]

By the end of the 1980s, it was becoming painfully clear that while the general population seemed to understand the *concept* of using the VCR to time shift broadcast television, the vast majority of VCR owners claimed an inability to actually do so. The first print reference to this sort of techno-logical incompetence appears in a 1988 review of a Los Angeles stand-up comedian, a "stereotypical football-playing, beer-swilling frat boy" who asked the crowd, "Any guys want to confess to me tonight they can't pro-gram the VCR? . . . If I want to record something that starts at 9 p.m., I can't leave the house until 9 p.m."[3]

The use of "I'm not able to program a VCR" as a figure of speech became more and more common in popular culture over the next few years. Such technological incompetence generally took one of two forms; either the inability to set the VCR's timer to tape future shows, or the more general inability to set the VCR's digital clock, with the latter possibility resulting in the oft-cited image of a blinking 12:00 as a tangible manifestation of the user's incompetence. In the late 1980s, some estimates placed the number of VCR owners who could not program their VCRs as high as 80 percent,[4] and this incompetence became a perceived sign of the vast distance between

technological experts and the nontechnological public. Remarked one col-
umnist, "You cannot program your VCR without a Ph.D. in electronics."[5] For
such users, programming a VCR involved recourse to external experts:
"Frequently I ran into the same problem despite several attempts to follow
what seemed to be clear, if complicated, directions," claimed one frustrated
owner, "but the tries led only to frustration, until a clerk from the store talked
me through the operation."[6]

By the early 1990s, the VCR owner who couldn't program his VCR was a
standard identity, drawn upon in a number of ways. In some cases, this image
of incompetence was used to humanize public figures, such as Ted Turner's
admission that he just hangs a washcloth over the blinking 12:00 on his bed-
room VCR.[7] This strategy also served to bring technological experts down to
a more approachable level; an article which began with the question "Are
[rocket scientists] the only ones whose VCR clocks don't flash 12:00 at them
for weeks?" concluded with a response by an actual rocket scientist that he
himself could only program his VCR with the help of the manual.[8] Along
these lines, the society page of the Arizona Republic reported:

When Joyce Downey was at Barbara Barrett's home for a Thelda Williams fund-raiser
Wednesday night, she noticed a large-screen television with a painting of a Pentium
processor above it. Underneath was the VCR with its telltale sign—a blinking clock.
"The rest of us don't need to feel so bad: Apparently, the head of Intel can't work his
either," said Downey, referring to Barbara's husband, Craig Barrett, the chief operating
officer of Intel.[9]

Other examples illustrate the ways in which an inability to program a VCR
became a part of a broader identity, that of feeling lost in the modern technological age. One essayist, reviewing the inventions that seemed to be displac-
ing her old manual typewriter, said "I can use the television set but not the
VCR,"[10] while a 1994 editorial claimed that "People who can't program a
VCR probably cringe at talk of an information superhighway."[11]

Those who still could program their VCRs, particularly those who worked
in the consumer electronics industry, often saw the problem as the fault of
the public; as one VCR salesman put it, "Most people are too lazy to open the
manual and do what they have to do."[12] VCR-related incompetence was even
associated with the declining scientific literacy of Americans, and scientists
decried "the numbers of scientifically illiterate adults who believe in astrolo-
gers and psychics and can't program their VCRs."[13] By 1999, writers were
linking VCR-related incompetence to broader technological hysteria:

... not everyone is as comfortable in this technological world as [some]. Many of us can't program our VCRs. Y2K technophobia has some folks storing food. Digital viruses are becoming more common than the influenza variety. And the computer world in many cases is the realm of hackers and spammers and pornographers ... oh my![14]

There is more to such rhetoric than simple incompetence, and these "discourses of incompetence" can function in similar ways to the "discourses of ignorance" used by workers in nuclear power plants to manage their nonscientific identities as well as their fears of radiation, as described by Mike Michael.[15] For my purposes here, however, the ways that such discourses function are not nearly as relevant as their sheer pervasiveness. Even a decade after Gemstar introduced VCRPlus, a system intended to make setting one's VCR a "non-intimidating, one-step process" involving entering a single number on one's remote control,[16] the VCR was still the popular choice for off-the-cuff riffs on hard-to-use technology.

How hard it actually was to program a VCR seems almost beside the point. Ironically, one survey showed that a little less than two-thirds of Americans were able to say "I fully understand how to operate my VCR and all of its features," far more than were able to say the same about their home security systems (45%), mobile phones (44%) or computers (44%).[17] Imagine, however, saying "I don't know how to use my cell phone to make a phone call" or "I can't use a computer mouse" to your friends and colleagues—that sort of talk would stigmatize a person, betraying a fundamental inability to use those technologies properly. The fact that it became socially safe to say that you could not program your VCR implied that programming your VCR was *no longer central* to its dominant technological frame, the watching of prerecorded movies and other programs in the home. The transformation begun by Andre Blay and the other pioneers of prerecorded video had reached its logical conclusion, so thoroughly recasting the VCR as a medium for movies that users' ability to use it for other purposes was quite literally atrophying.

Requiem for the VCR

As I write this, however, the VCR is a dying technology (and the description below will likely seem dated even by the time these words see print).[18] Though cheaper than ever, with VCRs retailing for as little as $30 apiece (a far cry from the first Betamax with its sticker price of $1,295 in 1976 dollars), new VCR sales are declining rapidly.[19] The three sometimes-overlapping functions of time shifting broadcast television, viewing homemade videos, and playing prerecorded movies have been usurped by other devices, leaving the VCR a shell of its former self.

On one hand, TiVo and other makers of digital video recorders (DVRs) are picking up the mantle of time shifting, explicitly framing their products in contrast with the seemingly inscrutable VCR. Though TiVo CEO Michael Ramsay claims that the set-top boxes are "revolutionizing TV,"[20] the DVR rhetoric of freedom and viewer empowerment bears a remarkable resemblance to the ad copy that accompanied Sony Betamax ads several decades earlier. The comparison with the VCR is oft-cited, though rarely flattering: "The best thing about this new service is how simple it is," explained one TiVo user. "It's not like the old VCR machines, which, let's be honest, none of us ever learned how to operate properly."[21] Moreover, time shifting as an action is increasingly being folded into the sociotechnical system of broadcast television, with cable companies building DVR technology directly into their cable boxes (and simultaneously incorporating this functionality into the technological frame of cable/satellite television).[22]

At the same time, the vast majority of video cameras sold today use either 8mm videocassettes or digital media, both of which are incompatible with traditional VCRs. Even older home movies on videocassette are looked down on, as consumers are advised that "[t]he tapes and the VCRs you play them on (which are getting increasingly scarce in electronics stores) do break down over time." The solution? "Convert your old videocassettes to DVDs—a far more handy and stable storage medium."[23]

Perhaps most importantly, the VCR is no longer the dominant technology for bringing movies into the home. In June 2003, less than six years after the debut of the format, weekly DVD rentals outpaced weekly videocassette rental for the first time, according to the Video Software Dealers Association.[24] Less than a year earlier, Circuit City (the second-largest electronics retail chain in the United States) had stopped selling movies on videocassette. Claimed a company spokesman: "We're responding to what people are wanting to buy."[25]

If videocassettes brought movies out of theaters and into the home, then DVDs took matters a step farther. A movie on DVD is embedded in a thick web of director commentaries, making-of documentaries, and production stills, all of which demystify The Movie itself, distracting from the movie-viewing experience and directing viewers' attention toward the circumstances of its production. DVDs tell viewers that they *should* in fact look at the man behind the curtain, and subtly shift the nature of movie watching from immersion in an experience to the abstracted analysis of a text. If industry insiders are to be believed, consumers overwhelmingly embrace this shift in orientation, a transformation that arguably owes a debt to the videocassettes that allowed everyday consumers to hold movies in their own hands for the first time.

Following the video stores who also adopted its garb, the world of home entertainment is saturated by the rhetoric of the movie theater. In the early years of the new millennium, virtually every home audio and video electronics manufacturer at the annual Consumer Electronics Show made reference to the idea of the "home theater," with the conventional wisdom that consumers wanted theaters (movie theaters, to be specific) in their homes unquestioned by all. Virtually every exhibitor demonstrated their television sets by playing Hollywood movies, manufacturers gave their product lines names like "Theatervision" and "Widescreen," and one couldn't walk ten feet without seeing the obligatory popcorn, flashing marquees, and red velvet ropes.

Ironically, the VCRs that were initially understood as accessories to the television set may well have played a part in the physical transformation of the television itself. At those CES shows, the screens of many newer television sets, particularly those following the High Definition Television (HDTV) standard, were framed not in the traditional 4:3 aspect ratio of NTSC television but rather the 16:9 widescreen aspect ratio of theatrical movies. Trumpeting its new and improved television sets, one manufacturer brought things full circle, going so far as to declare in bold, foot-high letters, "These TVs have been formatted to fit your movies."

Science & Technology Studies and Mediators

The overarching theme running throughout this story is the importance of mediation. On one level, the VCR's reconstruction from a time-shifting accessory to a movie machine for the home is the story of its definition as a medium. Movies had existed in other media before the VCR, but those media were always understood as "lossy"; somehow inferior to the traditional home for movies, the movie theater. When distributors created a new product by convincing studios to release their movies on videocassette, an opening was

Figure E.1
Photograph of the Philips Electronics booth, Consumer Electronics Show, Las Vegas, NV. (Photo by author, 2002.)

created, and the work of distributors, retailers, and clerks framed the new technology as tantamount to the movie theater, in the process redefining the nature of both the VCR and movies more generally. This reframing was made possible by the mediators who were not entrenched in or reliant upon preexisting understandings of home video, and who were able to persuade both producers and consumers of movies and technology to embrace this new technological frame.

Also saturating this analysis is an attention to the importance of mediation in knowledge production. If knowledge lies at the heart of any study of the social construction of science or technology,[26] then it seems important to follow that knowledge not just as it is created but as it is passed from one actor or one social group to another. Through its first decade, new channels of communication were established alongside new knowledge about the VCR and video stores. At times these channels were created from whole cloth, like the phone networks and newsletters of early videophiles, while others grew out of traditional relationships, such as retailer and customer, or distributor and retailer. If closure is a function of shared knowledge among relevant social groups, then the existence of stable communication chan-nels between those groups is a prerequisite to closure's establishment and maintenance.

In writing this history I have tried to foreground the efforts of those indi-viduals who themselves served as points of mediation, either by allowing a community to cohere through the shared exchange of knowledge or by facil-itating the exchange of knowledge across boundaries. Like the video distrib-utors who had to finesse the divergent interests of studios and retailers, such mediators often found themselves straddling multiple social groups, renego-tiating their identities based on their immediate context across the many strata between producer and consumer. Though their work is fluid and oft-overlooked, these mediators shaped that consumption junction in which consumers encountered a mass consumer technology, establishing a broader dialogue (and eventually, consensus) on its meaning.

This orientation toward mediation takes on added resonance when the artifact in question is used as a communication technology, and the informa-tion that it mediates is a particularly important aspect of the frame through which users understand it. In the case of the VCR, distributors and retailers didn't just define the technology's meaning, they offered an account of its relationship to the movies that it mediated, one which privileged the movie over the artifact (the cassette) on which it was encoded. These mediators drew a boundary around the prerecorded cassette, deliberately framing mov-ies on video in the context of *movies* rather than *video*. In the process, they established a new cultural meaning for the VCR that still reverberates today.

The greatest irony of this story is that distributors and retailers so firmly fixed this new understanding of both VCRs and movies in the cultural firma-ment that they ultimately rendered themselves irrelevant. As I remarked ear-lier, one of the essential characteristics of a mediator (or medium, for that matter) is that he or she exists in between two distinctly different entities. Were there no boundary, there would be no need for a mediator to straddle it. Moreover, the mediator is usually distinguished by her ability to present a comprehensible face to each side of the boundary she straddles.[27] Thus, a video distributor might pay a studio up front for cassettes that said studio saw as tangible artifacts to be paid for on delivery, while at the same time allowing video stores to treat those same cassettes like records or books, pay-ing for their weekly delivery once they'd begun to take in rental revenues. By in essence presenting one face to studios while offering a very different one to retailers, these distributors were able to finesse these very different (and incompatible) understandings of the product in question.

As time passed, however, and such mediators began to successfully bring varying actors around to a shared understanding of movies on video, the mediation they provided began to seem less essential to the smooth working of the consumption junction. In short, these obligatory passage points became unobligatory. In the case of movies on video, such irrelevance began to affect mediators at different levels of the distribution network; studios' home video divisions began to deal directly with retailers, larger superstores like Wal-Mart directly targeted the revenues of the video specialty retailers who had convinced customers to start buying videos (and later, DVDs) in the first place, and video clerks' expertise was challenged by the increasing sophistication of the very customers whom they themselves had educated. The implication is clear: if a mediator does a good enough job of evangelizing a shared frame to the various groups she connects, she risks rendering her unique mediating function unessential.

Perhaps the clearest example of this process in practice is the case of video stores themselves. In the early 1980s, it took a visionary with a certain appetite for risk to sink thousands of dollars into an inventory of tapes and open a video store. These retailers created a cultural institution, persuading their customers that this new consumption junction was an ideal place to spend their entertainment dollars. Once they had established the video store as a successful mediation space between Hollywood studios and consumers—accomplished by leveraging off the entrenched cultural meanings of the previous such mediation space, the movie theater—these retailers found themselves slowly being shaken out of the industry by larger conglomerates and franchises, whose strengths fell more in the realm of efficiencies of scale than in cultural innovation. While a handful of stores managed to survive, most found that their particular brand of mediation could no longer bring in enough revenues and either cashed out or folded.

At the risk of dating this book, things have truly come full circle by 2006, with Blockbuster and the handful of other large-scale franchise chains facing a substantial threat from Netflix and other DVD-by-mail services. Founded in 1997, Netflix offers customers the ability to "rent" as many movies as they like for a flat monthly fee: under their standard plan, a subscriber creates a list of desired titles on the Netflix website and is sent the first three DVDs by mail. Once she has watched one, she returns it by mail and is sent the next DVD in her queue that day. Ironically, in the same way as the video store originally co-opted the rhetoric of the movie theater, presenting itself as "just like the theater, but better," Netflix is succeeding by framing itself as "just like a video store, but better." Its founder, Reed Hastings, is fond of telling an origin story that pins Netflix's entire raison d'etre on a $40 fee he received for returning a copy of *Apollo 13* late to his local video store, and the Netflix website touches on many of the independent video store's tropes, from the "unique and personal movie recommendations" that it offers (recalling the heyday of the video clerk) to the diverse library of DVDs available for rental. At the same time, however, Netflix positions itself as an improvement on the brick-and-mortar video store, from the freedom from late fees that it offers to its eco-friendly claim that "If Netflix members, instead of receiving movies by mail, drove two miles each way to a rental store, they would consume 250,000 gallons of gasoline per day and release 750,000 tons of carbon dioxide emissions annually."[28] As Blockbuster and Wal-Mart, the titans of brick-and-mortar video rental and sell-through (respectively), both try to navigate the DVD-by-mail business, it's worth keeping in mind that the very cultural meanings of movies and video stores that they are trying to make seem old-fashioned and out-of-date were, in fact, created by earlier retailers in response to previous cultural meanings.

As for the medium itself, the DVD represents an attempt by Hollywood studios to reclaim the control of their products that they lost to mediators and consumers in the 1980s. Unlike videocassettes, movies on DVD are encoded with a proprietary algorithm called Copyright Scrambling System (CSS), which is created, owned, and regulated by the DVD Copy Control Association (DVD CCA), a consortium of movie studios and hardware manufacturers.[29] Since studios only release DVD movies that are encrypted with CSS, they can essentially control the format through the DVD CCA. Motion picture studios have found other ways to even more explicitly control the distribution and sale of movies on DVD; for example, the DVD standard divides the world into seven regions, and a disc produced for one region will not play in a DVD player designed for another. In a sense, the DVD standard represents the motion picture studios' attempt to learn from their experiences with the VCR, particularly their powerlessness in the face of the mediators who sat at the heart of the industry. By inserting control mechanisms like CSS and region encoding into the technical standard of the format itself, the studios have rewritten the relationship between themselves and both mediators and the new medium itself.[30]

Mediation Matters

If there is a lesson to be learned from the story of the video store, it may simply be that mediation matters. On one hand, the history of a consumer technology is not simply a story of producers and consumers, but also of the varying levels of mediation that lie between the two. The consumption junction is a valuable device for understanding the choices offered to consumers, but in order to understand how a specific configuration of the consumption junction comes to be, we must look beyond the consumer's perspective to the actions of the mediators, who create a context for manufacturers' products. Without doing so, we wouldn't know why most movies on videocassette were rented rather than sold, why the video store included videogames but not record players, or why the clerk behind the counter couldn't fix a VCR but could name every film directed by Alfred Hitchcock. Mediators may not have had much influence on the material nature of their wares, but they did shape the material spaces in which those artifacts were situated and offered to consumers, in the process literally constructing the consumption junction.

When the technology in question is used for communication, mediation takes on added significance as an integral part of its technological frame. *By definition*, a communication technology mediates information, since it is the construction of that information (and the boundary between it and the artifact) that marks the technology as a medium. Once we understand this unique property of a communication technology, we see that the history of media isn't just a story of senders and receivers, but of how both understand the channels that connect them. Without an awareness of the coproduced nature of medium and message, we might overlook the fact that the VCR's redefinition as a "movie machine" entailed a redefinition not just of the recorder, but also of movies themselves. This perspective opens up both medium *and* message to the same kind of analysis, one focusing not on the inherent properties of either, but rather on the ways that interrelated knowledge is produced about both.

Ultimately, there is no such thing as truly invisible mediation, but at the same time mediators and media don't have a purely deterministic role; they are merely one more piece of a given sociotechnical network, another component mutually shaping the others. Often overlooked in favor of the entities between which they sit, mediators and media serve an essential function, offering a unique means of understanding both information and technology. By focusing our scrutiny on the people and technologies in between, we can better understand the nature of the spaces they connect and help to define.

Appendix: Notes on the videostoreproject.com Website

Though most of the material on which this book draws was gathered via traditional archival research methods and in-person or telephone interviews, a distinct and important part of the research involved the collection of oral histories using the Internet. While quite useful, the stories and memories accumulated via the World Wide Web do raise useful methodological questions, which are worth addressing here.

Background

In February 2002, I was invited to take part in an Exploring and Collecting History Online (ECHO) Project workshop hosted by the Center for History and New Media (CHNM) at George Mason University. Over the course of the two-day workshop, I was introduced to a variety of projects that had used the Web to collect materials on various events and figures in the history of science and technology, from polymerase chain reaction and the computer mouse to the East Coast blackouts of the 1970s.[1] At around the same time, I was trying to figure out how one might gather materials on a widespread yet distributed phenomenon like the early history of home video. While I'd already begun interviewing prominent storeowners and other obvious key figures, it wasn't clear how I could gather information on the perspective of everyday customers in the video stores of the early 1980s—while newspaper articles and advertising materials offered one glimpse into these stores, there were remarkably few historical traces of the actual experience of visiting one. Simply asking everyone I met for their video store memories helped a bit, but the sample was limited both by time and geographic constraints. It made sense that a Web-based collection tool might be able to extend my reach, allowing me to cast a net for stories of early video stores beyond my own limited social network. By late 2002, I had built what I called the "Video Store Project" using the open source scripting language PHP and the open source database server mySQL (for additional flexibility, I programmed the site from scratch rather than using ECHO's Survey Builder). The site was hosted on the CHNM server, but for ease of access I registered the domain name videostoreproject.com (at which address it still resides).

When opening the site, a user is offered the choice of telling their story or reading the memories of others. If contributing, the respondent is first asked whether he or she owned a video store, worked in a video store, or simply shopped in a video store. (I deliberately refrained from defining exactly what a video store was, to avoid prejudicing any responses.) Based on the response, the site asks a series of questions about the respondent's experiences, from the physical layout of the store and the kinds of merchandise it stocked to the relationships between the people on opposite sides of the counter.

Naturally, I was concerned with the privacy of my respondents. While I did ask for a name and e-mail address for verification purposes, responses are effectively anonymous, as no identifying information is publicly available on the website. A proposal was also submitted to the Cornell Human Subjects Review Board, which declared the project exempt from review.[2] In the course of running the site, I only had one experience in which a respondent directly contacted me regarding her anonymity—a former store employee who had critiqued her ex-employer only to later find that he had responded to the survey as well (she asked me to take down her response from the public site and I complied).

In designing the survey itself, I deliberately spread the questions over several pages so that a visitor wouldn't be confronted with an interminable form. My assumption was that a person would be more likely to answer a few questions than a dozen, and that having invested an initial amount of time in a response, would be more likely to continue answering questions on subsequent pages. It's not clear how accurate this assumption was—around 45 percent of respondents filled out the first page of the survey but didn't complete the rest.[3]

Responses

One of the strongest lessons of the ECHO workshop I attended was that the most successful oral history websites were those that tapped into a deeper group experience, one that inspired members to spend their time filling out an online survey. As it turned out, the early days of home video seemed to inspire a potent mixture of enthusiasm and nostalgia, reflected in the site's ultimate success.

Initially, I publicized the site via a mass e-mail to friends, colleagues, and family, encouraging them to forward the announcement to others. The word spread virally, and by early March the site had logged 130 visitors. I also posted notices to relevant Usenet and other discussion groups, publicizing the site further (albeit drawing the ire of several participants who accused me of violating "netiquette" by posting a survey). In addition, I reached out to the video industry through trade publications and organizations, and the site was profiled in a story on *Video Store Magazine*'s website.[4] The real break-through happened on March 6, 2003, when my e-mail request was posted on *Ain't It Cool News*, a leading film gossip website with a large and vocal community.[5] By the end of that day, the Video Store Project logged more than 2,000 unique visitors, about a quarter of whom filled out a survey. Over the next few days, the torrent slowed, but as of December 2006 the site is still collecting responses, with the total number around 1,275 (of which roughly 625 can be considered substantially incomplete).

As of the time of this writing, the site has collected 178 surveys from store-owners, 275 surveys from employees and 833 surveys from customers, with the average respondent spending just over twenty eight minutes to fill out a complete survey. As one might imagine, the responses range from incredibly verbose to incredibly terse. At their best, the survey responses are valuable snapshots of a lost moment in time, describing life in both urban and rural video stores across the country. They tend to be vibrant (at times verging on affectionate) recollections of a now-mundane experience that, at the time, was positively thrilling. The responses overall are surprisingly earnest, and many echo the sentiments of one visitor: "Thanks . . . this sounds strange but you made me think about something I haven't thought about in a long time . . . brought back a lot of good memories."[6]

As a researcher, however, this material does raise important questions above and beyond the standard concerns of survey-based questioning. For one, the interview sample is entirely self-selecting; in order to be included, a respondent had to come across the site one way or another and explicit-ly choose to fill out a survey, with the consequence that visitors who didn't have any strong feelings about the early days of home video simply didn't bother to respond. Thus, the responses naturally trend toward the extremes, both positive and negative, and relatively neutral sentiments, no matter how widespread, tend not to show up in the results. Moreover, while virtually anybody can find access to a computer using a public library, it seems a pretty safe bet that the results are skewed by socioeconomic class and other factors that predict computer ownership and Internet access.

There is also a question of reliability—as the oft-cited New Yorker cartoon might imply, on the Internet, nobody knows if the survey's being filled out by a dog. A handful of the responses are obviously in bad faith (one standout is written from the perspective of the protagonist of Kevin Smith's movie *Clerks*), but for the most part the stories seemed genuine. I generally took the survey responses in good faith, but when citing a specific anecdote, I did make every effort to contact the respondent in order to verify both his or her identity and the details of the story.

Ultimately, I wouldn't argue that one can base broad conclusions on the results of an online survey like the Video Store Project, and I didn't try to do so in this research. At no point does my argument hinge on evidence from the website. Instead, I use the responses to flesh out my narrative, to lend additional detail and color to the conclusions that I'd reached through a combination of interviews and archival research. In addition, I found that the website itself was an invaluable tool that augmented my in-person interviewing by expanding my pool of interviewees beyond the core set I was able to identify through trade publications and snowball interviewing—I never would have met Michael Dark, Henry Tolino, or Mike Salomon, among many others, had I not followed up their survey response with an e-mail or phone call. Ultimately, the Video Store Project offered a valuable addition to the other methods in my researcher's toolkit, one for which this narrative is substantially richer.

Notes

Introduction

1. Frederick Wasser, *Veni, Vidi, Video* (Austin: University of Texas Press, 2001), 4.

2. For more on this particular use of the VCR, see James M. Moran, *There's No Place Like Home Video* (Minneapolis: University of Minnesota Press, 2002) and Lawrence J. Vale, "Captured on Videotape: Camcorders and the Personalization of Television," in *Social and Cultural Aspects of VCR Use*, ed. Julia R. Dobrow (Hillsdale, NJ: Lawrence Erlbaum Associates, 1990).

3. For a general overview of the sociological literature on VCR use, see Julia R. Dobrow, ed., *Social and Cultural Aspects of VCR Use* (Hillsdale, NJ: Lawrence Erlbaum Associates, 1990); Ann Gray, *Video Playtime: The Gendering of a Leisure Technology* (London: Routledge, 1992); Mark R. Levy, ed., *The VCR Age* (Newbury Park: SAGE Publications, 1989).

4. To a contemporary analyst, Universal's reaction might seem counterintuitive, if not downright foolhardy. One would think that the VCR would be welcomed as a means of cultivating a larger audience for both shows (allowing viewers to essentially watch two programs at the same time). Seen in this light, Universal's heavy-handed response throws into sharp relief the extent to which they thought of their programming as a commodity whose value lay at least in part in its scarcity.

5. James Lardner, *Fast Forward: Hollywood, the Japanese, and the Onslaught of the VCR* (New York: Norton, 1987).

6. *Universal v. Sony* (1976).

7. Margaret Graham offers a related account of RCA's earlier experiments with the VideoDisc as an alternate standard for home movie distribution. Margaret B.W. Graham, *RCA and the Videodisc* (Cambridge: Cambridge University Press, 1986).

8. Lardner, *Fast Forward*, 161.

9. See, respectively: Ibid.; Eugene Marlow and Eugene Secunda, *Shifting Time and Space. The Story of Videotape* (New York: Praeger, 1991); P. Ranganath Nayak and John M. Ketteringham, *Breakthroughs!* (Amsterdam: Pfeiffer & Company, 1994); Wasser, *Veni, Vidi, Video*.

10. M. Cusumano, Y. Mylonadis, and R. Rosenbloom, "Strategic Maneuvering and Mass-Market Dynamics—The Triumph of VHS over Beta," *Business History Review* 66, no. 1 (1992): 86.

11. Wasser, *Veni, Vidi, Video*, 74.

12. Jonathan Coopersmith, "Pornography, Technology, and Progress," *ICON* 4 (1998): 97–125.

13. Marc Berman, "See No Evil," *Video Business*, November 1984.

14. Drucilla Cornell, ed., *Feminism and Pornography, Oxford Readings in Feminism* (Oxford: Oxford University Press, 2000); Susan Dwyer, *The Problem of Pornography* (Belmont, CA: Wadsworth, 1995); Catharine MacKinnon, *Pornography and Sexual Violence: Evidence of the Links* (London: Easton, 1988); Linda Williams, *Hard Core: Power, Pleasure, and the "Frenzy of the Visible"* (Los Angeles: University of California Press, 1999).

15. One might find an interesting resonance here with the relationship between actors and actants described by proponents of actor-network theory.

16. Bruno Latour, *Science in Action: How to Follow Scientists and Engineers through Society* (Cambridge, MA: Harvard University Press, 1987).

17. Gail DeGeorge, *The Making of a Blockbuster* (New York: John Wiley & Sons, 1996).

18. Ruth Schwarz Cowan, "The Consumption Junction: A Proposal for Research Strategies in the Sociology of Technology," in *The Social Construction of Technological Systems*, ed. Trevor Pinch, Wiebe Bijker, and Thomas Hughes (Cambridge, MA: MIT Press, 1987).

19. As I'll discuss later in chapter 2, there have been few studies of such mediators in the Technology Studies literature, and most of those focus on mediators like home economists and sales agents who, rather than working as free agents, are explicitly employed by the producers of the technologies and knowledge that they mediate. See Carolyn Goldstein, "From Service to Sales: Home Economics in Light and Power, 1920–1940," *Technology and Culture* 38, no. 1 (1997); Ronald R. Kline, *Consumers in the Country: Technology and Social Change in Rural America* (Baltimore: Johns Hopkins University Press, 2000); Trevor Pinch and Frank Trocco, *Analog Days: The Invention and Impact of the Moog Synthesizer* (Cambridge, MA: Harvard University Press, 2002); Oliver Zunz, *Making America Corporate, 1870–1920* (Chicago: University of Chicago Press, 1990).

20. Graham, *RCA and the Videodisc*.

21. On the life of the videocassette recorder beyond North America, see Manuel Alvarado, ed., *Video Worldwide* (Paris: United Nations Educational, Scientific, and Cultural Organization, 1988) and Mark R. Levy, ed., *The VCR Age. Home Video and Mass Communication* (Newbury Park: Sage Publications, 1989), especially these included articles: Douglas A. Boyd, "The Videocassette Recorder in the USSR and Soviet-Bloc countries; Barrie Gunter and Thomas R. Lindlof, "The Uses and Impact of Home Video in Great Britain"; Christine Ogan, "The Worldwide Cultural and Economic Impact of the VCR"; Dov Shinar, "VCR Narrowcasting in the Kibbutz."

22. C. E. Shannon and W. C. Weaver, *The Mathematical Model of Communication* (Urbana: University of Illinois Press, 1949).

23. The point is made even more firmly by the fact that it's easy to imagine a scenario in which that same user might come to believe that the hammer *is* in fact conveying information—nothing has changed but the frame through which the hammer-watcher is viewing the hammer, but it has immediately become a medium in her eyes.

24. For a detailed examination of this research tradition, see Daniel J. Czitrom, *Media and the American Mind: From Morse to McLuhan* (Chapel Hill: University of North Carolina Press, 1982), Everett M. Rogers, *A History of Communication Study: A Biographical Approach* (New York: The Free Press, 1994).

25. Jay G. Blumler and Elihu Katz, *The Uses of Mass Communication: Current Perspectives on Gratifications Research* (Newbury Park, CA: Sage, 1974).

26. Nicholas Abercrombie and Brian Longhurst, *Audiences* (London: SAGE Publications, 1998); Ien Ang, *Desperately Seeking the Audience* (London: Routledge, 1991); James W. Carey, *Communication as Culture: Essays on Media and Society* (Boston: Unwin Hyman, 1989); Virginia Nightingale, *Studying Audiences: The Shock of the Real* (London: Routledge, 1996); John Tulloch, *Watching Television Audiences* (New York: Oxford University Press, 2000).

27. Susan J. Douglas, *Inventing American Broadcasting, 1899–1922* (Baltimore: Johns Hopkins University Press, 1987); Susan Smuylan, *Selling Radio: The Commercialization of American Broadcasting, 1920–1934* (Washington, DC: Smithsonian Institution Press, 1994).

28. Gregory J. Downey, *Telegraph Messenger Boys: Labor, Technology, and Geography, 1850–1950* (New York: Routledge, 2002); Carolyn Marvin, *When Old Technologies Were New: Thinking About Electric Communication in the Late Nineteenth Century* (New York: Oxford University Press, 1988).

29. Claude S. Fischer, *America Calling—A Social History of the Telephone to 1940* (Berkeley, CA: University of California Press, 1992); Michèle Martin, *"Hello, Central?":*

Gender, Technology, and Culture in the Formation of Telephone Systems (Montreal: McGill-Queen's University Press, 1991).

30. Cecelia Tichi, *Electronic Hearth: Creating an American Television Culture* (New York: Oxford University Press, 1991).

31. Andre J. Millard, *America on Record: A History of Recorded Sound* (Cambridge: Cambridge University Press, 1995); David Morton, *Off the Record: The Technology and Culture of Sound Recording in America* (New Brunswick, NJ: Rutgers University Press, 2000).

32. Janet Abbate, *Inventing the Internet* (Cambridge, MA: MIT Press, 1999); Sherry Turkle, *Life on the Screen: Identity in the Age of the Internet*, 1st Touchstone ed. (New York: Simon & Schuster, 1997).

33. Favorite examples include: Susan J. Douglas, *Where the Girls Are: Growing up Female with the Mass Media*, 1st pbk. ed. (New York: Times Books, 1995); Michele Hilmes, *Radio Voices: American Broadcasting, 1922–1952* (Minneapolis, MN: University of Minnesota Press, 1997); Jeffrey Sconce, *Haunted Media* (Durham, NC: Duke University Press, 2000); Robert Sklar, *Movie-Made America: A Cultural History of American Movies*, rev. ed. (New York: Vintage Books, 1994); Lynn Spigel, *Make Room for TV: Television and the Family Ideal in Postwar America* (Chicago: University of Chicago Press, 1992).

34. The best examples of such balance are Susan Smuylan's *Selling Radio*, which charts the construction of both the technology and its advertising conventions, and Susan Douglas's books *Listening In* and *Inventing American Broadcasting* which, when taken together, offer a comprehensive survey of the history of radio technology and content (Smuylan, *Selling Radio*; Douglas, *Inventing American Broadcasting, 1899–1922* and *Listening In: Radio and the American Imagination, from Amos 'N' Andy and Edward R. Murrow to Wolfman Jack and Howard Stern*, 1st ed. (New York: Times Books, 1999)).

35. Marshall McLuhan, *Understanding Media: The Extensions of Man*, 1st MIT Press ed. (Cambridge, MA: MIT Press, 1994).

36. Neil Postman, *Amusing Ourselves to Death* (New York: Penguin Books, 1985); Neil Postman, *Building a Bridge to the 18th Century* (New York: Knopf, 1999); Neil Postman, *Technopoly: The Surrender of Culture to Technology* (New York: Vintage Books, 1993).

37. Joshua Meyrowitz, *No Sense of Place: The Impact of Electronic Media on Social Behavior* (New York: Oxford University Press, 1985).

38. Paul Levinson, *The Soft Edge: A Natural History and Future of the Information Revolution* (London: New York: Routledge, 1997).

39. To be more precise, Bijker lists among the components of a technological frame "goals, key problems, problem-solving strategies, requirements to be met by problem solutions, current theories, tacit knowledge, testing procedures, design methods and

criteria, users' practice, perceived substitution function, and exemplary artifacts" (Wiebe E. Bijker, *Of Bicycles, Bakelites, and Bulbs: Toward a Theory of Sociotechnical Change* (Cambridge, MA: MIT Press, 1995), 122–125).

40. I'm simply focusing on media technologies here, but they are actually a part of a larger family of technologies that one might call "information" technologies, all of which share the structural similarity of a boundary between their material nature and their immaterial *information*. One can situate, for example, technologies of communication (which move information from place to place), of processing (which transform information according to specific rules), and of storage (which allow information to persist over time) under this broader umbrella.

41. One of the more interesting analyses of this transcendence of the message is Sconce's *Haunted Media*, which offers a cultural history of ideas of "presence" in electronic media throughout the late nineteenth and twentieth centuries. A common discourse about electronic media invokes the supernatural, with media (particularly in the initial phases of their introduction) being characterized as "haunted" with "ghostly" or otherwise supernatural presences. Sconce situates media such as the telegraph and television within this discourse, explaining the rhetoric of presence that electronic media seem to have inspired and elaborating on the often explicit connections made by historical actors between new electronic media and spiritual mediums, extraterrestrial contact, and alternate realms of existence (Sconce, *Haunted Media*).

42. While one can argue that a toddler watching *Teletubbies* doesn't have this sophisticated understanding that the images on screen are being sent from elsewhere, it seems fair to say that such a case represents deliberate "use" of the television about as much as a hamster can be said to be "using" the wheel on which he runs.

43. Bruno Latour, *We Have Never Been Modern* (Cambridge, MA: Harvard University Press, 1993).

44. This idea bears a structural similarity to the ways in which a technology designer constructs an ideal user, a construction which then informs the design process. See Tarleton Gillespie, "The Stories Digital Tools Tell," in *New Media: Theories and Practices of Digitextuality*, ed. Anna Everett and John T. Caldwell (New York: Routledge, 2003); Julian Albert Kilker, "Networking Identity: A Case Study Examining Social Interactions and Identity in the Early Development of E-Mail Technology" (doctoral dissertation, Cornell University, 1999); Christina Lindsay, "Invisible Computers and Constructed Users: The TRS-80 Computer 20 Years On" (paper presented at the Technology & Identity conference, Cornell University, 1999); Jameson Wetmore, "Systems of Restraint: Redistributing Responsibilities for Automobile Safety in the United States since the 1960s" (doctoral dissertation, Cornell University, 2003).

45. Susan Leigh Star and James Griesemer, "Institutional Ecology, 'Translations,' and Boundary Objects: Amateurs and Professionals in Berkeley's Museum of Vertebrate Zoology, 1907–1939," *Social Studies of Science* 19 (1989): 387–420.

46. This is a transformation that has been essentially taken for granted in scholarship on both film and technology. The best analysis thus far is Frederick Wasser's *Veni Vidi Video*, but Wasser is less interested in how movies got onto video than he is in the consequences of that process, using it to explain the film industry's trend toward the production of larger and larger blockbuster movies in recent years. The rest of the literature verges on the teleological, describing the emergence of movies on video as a natural consequence of the technology of the VCR.

Chapter 1

1. Ray Glasser, *Video Collectors of Ohio 1979 Convention* (Fremont, OH: 1979), videotape.

2. Art Vuolo, "Michigan Update," *The Videophile's Newsletter*, September/October 1978.

3. Marc Wielage and Rod Woodcock, "From the Editors," *Videofax: The Videophile's Newsletter*, Fall 1983.

4. For a remarkable survey of movies in the home before video, see Alan Kattelle, *Home Movies: A History of the American Industry, 1897–1979* (Nashua, NH: Transition Publishing, 2000).

5. Anthony Slide, *Before Video: A History of the Non-Theatrical Film*, vol. 35, *Contributions to the Study of Mass Media and Communications* (New York: Greenwood Press, 1992), 115–116.

6. Bill Farley (vice president of play boy Enterprises), e-mail to the author, April 12, 2004; Marc Wielage and Rod Woodcock (authors of "The Rise and Fall of Beta"), interview with the author, March 15, 2003.

7. Sammy Davis Jr. owned a more portable version of Hefner's setup, bringing a cart of video equipment every time he checked into a particular Las Vegas hotel so that he could tape soap operas off TV during the day and watch them after his evening performances. Michael Dark (early video retailer), interview with the author, March 31, 2003.

8. Joe Mazzini, "3/4" U-Matic Exchange Notes," *The Videophile's Newsletter*, October 1976.

9. Marc Wielage and Rod Woodcock, interview.

10. Ibid.

11. Ray Glasser, "Reflections on Owning a Betamax," *The Videophile's Newsletter*, July/August 1977; Jim Lowe (*The Videophile's Newsletter*), interview with the author, March 7, 2003.

12. Sony Betamax Advertisement, *TV Guide*, October 1, 1977.

13. Sony Betamax Advertisement, *TV Guide*, December 2, 1977.

14. Sony Betamax Advertisement, *TV Guide*, October 22, 1977.

15. Needless to say, no mention is made of the hours one might lose from another day in order to watch recorded programs (Sony Betamax Advertisement, *TV Guide*, June 10, 1978).

16. According to Lardner (*Fast Forward: Hollywood, the Japanese, and the Onslaught of the VCR*, pp. 21–22), the first time that the president of Universal Pictures even heard of the Betamax was when Sony asked for permission to mention *Kojak* in an advertisement.

17. As I'll discuss later, the adult industry showed more flexibility in adapting to this new medium, and established a prerecorded market well in advance of more main-stream entertainment. As for the illegal market for first-run entertainment, I'll leave that for the next chapter.

18. I'm using the masculine pronoun here and throughout this chapter because these hobbyists were overwhelmingly male.

19. Jim Lowe, *The Videophile's Newsletter*, September 1976, 4.

20. George Mair, *Inside HBO: The Billion Dollar War between HBO, Hollywood, and the Home Video Revolution* (New York: Dodd, Mead & Company, 1988); "TV from Space Available at Home," *Home Satellite TV News*.

21. Jim Lowe, *The Videophile's Newsletter*, September 1976, 6.

22. Michael Dark, interview.

23. Jonathan (last name unknown), online survey filed at The Video Store Project web-site, March 6, 2003, available from http://echo.gmu.edu/workshops/jgreenberg/viewdetail.php?responseID=270.

24. Melanie March, online survey filed at The Video Store Project web-site, March 7, 2003, available from http://echo.gmu.edu/workshops/jgreenberg/viewdetail.php?responseID=651.

25. Dark, interview.

26. Brad Burnside (chairman of Sourcelight Technologies, Inc.), interview with the author, June 4, 2003.

27. Ray Glasser (of the Ultimate Betamax Information Guide), interview with the author, May 16, 2003.

28. Ibid.

29. Jim Lowe, *The Videophile's Newsletter*, September 1976, 2.

30. Lowe, interview.

31. Ibid.

32. Glasser, "Reflections on Owning a Betamax," 9.

33. Glasser, interview.

34. Ibid; Cassette tape from the private collection of Ray Glasser.

35. Wielage and Woodcock, interview.

36. Ray Glasser, "How to Do It! Putting on a Video Convention!," *The Videophile*, September/October 1979.

37. Jim Lowe, *The Videophile's Newsletter*, September 1976, 2.

38. Jim Lowe, *The Videophile's Newsletter*, December 1976, 8.

39. Glasser, interview.

40. Jim Lowe, *The Videophile's Newsletter*, December 1976, 3.

41. Glasser, interview; Mark Stencel (former video store clerk), interview January 31, 2003; Wielage and Woodcock, interview.

42. Jim Lowe, *The Videophile's Newsletter*, December 1976, 9.

43. Though in a different context, the ways in which early videophiles spoke of having a "pretty skilled 'touch'" recall Harry Collins' descriptions of the ways in which tacit knowledge is passed among scientists working with lab equipment. H. M. Collins, *Changing Order: Replication and Induction in Scientific Practice* (London: SAGE Publications, 1985).

44. Marc Wielage, "The Videophile Product Report: The Editor/Muntz Commercial Cutter," *The Videophile*, December 1979.

45. Letter from "Orlando videophile," published in Jim Lowe, *The Videophile's Newsletter*, January/February 1977, 10.

46. Stencel, interview.

47. Letter from "my main man in Southern California," published in Jim Lowe, *The Videophile's Newsletter*, January/February 197; 10.

48. Ibid, 11.

49. Among the original broadcast tapes in Glasser's collection, he's most proud of "two film chain copies of Star Trek from the 60s with original commercials" (Glasser, interview).

50. Jim Lowe, *The Videophile's Newsletter*, December 1976, 4.

51. Wielage and Woodcock, interview.

52. Jim Lowe, "TV Wiggles," *The Videophile's Newsletter*, March/April 1978; Joe Mazzini, "Joe Mazzini's U-Matic and Beta Notes," *The Videophile's Newsletter*, March/April 1978; Art Vuolo, "State of the Art," *The Videophile's Newsletter*, January/February 1978.

53. Jim Lowe, "Hooking Up," *The Videophile's Newsletter*, March/April 1978.

54. Lothar Merker, "Letter to the Editor," *The Videophile*, July/August 1979.

55. Redoutey and *The Videophile's Newsletter* contributor Art Vuolo later made appearances together on radio and television talk shows (Art Vuolo, "State of the Art," *The Videophile's Newsletter*, September/October 1979; Art Vuolo, "State of the Art," *The Videophile*, September 1981).

56. Charlie Lull, "Letter to the Editor," *The Videophile's Newsletter*, July/August 1979.

57. Jim Lowe, *The Videophile's Newsletter*, December 1976, 9.

58. Marc Wielage, for example, remembers having to tell his local electronics retailer that Sony was even going to market a stand-alone VCR in the first place: "I talked to the biggest local Sony dealer in town . . . and they said 'We don't think Sony will ever market it separately. We think that they're marketing VCRs as a way to sell TV sets.' I said, 'I think you're wrong. I think they're going to market a standalone VCR.' They went, 'Oh, what do you know?" Well I was able to come back a month or two later . . . [after] an announcement on the front page of Billboard magazine that [Sony] was going to sell a standalone VCR deck for the United States. I went into the store and showed them [the advertisement], and they were absolutely blown away, because even though they were the biggest Sony dealer in central Florida, even Sony had not told them that they were going to do that" (Wielage and Woodcock, interview).

59. Joe Mazzini, "3/4" U-Matic Exchange Notes," *The Videophile's Newsletter*, December, 197

60. Susan J. Douglas, *Inventing American Broadcasting, 1899–1922* (Baltimore: Johns Hopkins University Press, 1987), chapter 6.

61. On dx-ers, see Susan J. Douglas, *Listening In: Radio and the American Imagination, from Amos 'N' Andy and Edward R. Murrow to Wolfman Jack and Howard Stern* (New York: Tim Books, 1999), chapter 3. On later radio hobbyists, see Kristen Haring, *Ham Radio's Technical Culture*, (Cambridge, MA: MIT Press, 2006); Yuzo Takahashi, "A Network of Tinkerers: The Advent of the Radio and Television Receiver Industry in Japan," *Technology and Culture* 41, no. 3 (2000).

62. Joseph O'Connell, "The Fine-Tuning of a Golden Ear: High End Audio and the Evolutionary Model of Technology," *Technology and Culture* 33, no. 1 (1992).

63. For more on personal computer hobbyists, see Martin Campbell-Kelly and William Aspray, *Computer: A History of the Information Machine* (New York: Basic Books, 1996), chapter 10; Steven Levy, *Hackers* (New York: Dell Publishing, 1984), part two.

64. On early automobile users, see Ronald Kline and Trevor Pinch, "Users as Agents of Technological Change: The Social Construction of the Automobile in the Rural United States," *Technology and Culture* 37 (1996). For more on hot rod culture, see John F. DeWitt, *Cool Cars, High Art: The Rise of Kustom Kulture* (Jackson, MS: University Press of Mississippi, 2002); Robert C. Post, *High Performance: The Culture and Technology of Drag Racing 1950–1990* (Baltimore: Johns Hopkins University Press, 1994).

65. Haring, *Ham Radio's Technical Culture*, 10.

66. For more on such materials, see Takahashi's excellent analysis of the radio and television tinkering community in Japan, and Théberge's discussion of the electronic music enthusiast community's activities (Takahashi, "A Network of Tinkerers: The Advent of the Radio and Television Receiver Industry in Japan," Paul Théberge, *Any Sound You Can Imagine* (Hanover, NH: Wesleyan University Press, 1997).

67. Levy, *Hackers*, 7.

68. Sherry Turkle, *The Second Self: Computers and the Human Spirit* (New York: Simon and Schuster, 1984), 225.

69. Levy, *Hackers*, 102–103.

70. For an introduction to the concept of the sociotechnical system, see Thomas P. Hughes, "The Evolution of Large Technological Systems," in *The Social Construction of Technological Systems*, ed. Wiebe Bijker, Thomas Hughes, and Trevor Pinch (Cambridge, Massachussetts: The MIT Press, 1987).

71. Turkle, *The Second Self*, 225.

72. Douglas, *Listening In*; Brooke Hindle, *Emulation and Invention* (New York: Basic Books, 1981).

73. Ron Rosenbaum, "Secrets of the Little Blue Box," *Esquire*, October 1971.

74. Helen Nissenbaum, "Hackers and the Contested Ontology of Cyberspace," *New Media & Society* 6, no. 2 (2004).

75. The story of this case has been well told elsewhere, particularly in Lardner's *Fast Forward: Hollywood, the Japanese, and the Onslaught of the VCR*.

76. Jim Lowe, The Videophile's Newsletter, March/April 1978, 3–8.

77. Marc Wielage, "Rambling Outtakes," *The Videophile*, March/April 1979. While on the stand, Marc had the dubious honor of playing a segment of *The Mummy's*

Hand in order to demonstrate how he edited the commercials out of broadcast movies.

78. *Universal v. Sony.*

79. Jim Lowe, "A Bald-Faced Plea," *The Videophile's Newsletter*, 480 F. Supp. 429 (1979). September/October 1978.

80. Lowe, interview.

81. Lowe, "TV Wiggles"; Vuolo, "State of the Art."

82. The CES booth is notable because at that time, enthusiasts who weren't officially working in the electronics industry weren't allowed inside the show—thus, *The Videophile* was targeting the industry more than reaching out to its readers (even though the two groups often overlapped).

83. Wielage and Woodcock, "From the Editors," *Videofax: The Videophile's Newsletter*, Fall 1983, 2–4.

84. Wielage and Woodcock, interview.

85. Glasser, interview.

Chapter 2

1. For a detailed examination of why jokes are funny, see M. Mulkay, *On Humour* (Cambridge: Polity Press, 1984).

2. This section relies heavily on Nmungwun's detailed history of both nonmagnetic and magnetic video recording technologies (Aaron Foisi Nmungwun, *Video Recording Technology: Its Impact on Media and Home Entertainment* (Hillsdale, NJ: Lawrence Erlbaum Associates, 1989).

3. Though the initial prototype required a user to insert a finger and thread the tape around the playback head manually, the design was quickly revised to include a carousel mechanism that needed no such skilled intervention by the user. It was at this time that the machine was dubbed the "U-Matic," in a reference to this loading mechanism (Peter Keane (Former Sony spokesman), interview with the author), January 6, 2004.

4. Ibid.

5. Nmungwun, *Video Recording Technology*. Chapter 6.

6. Jonathan Coopersmith, "Pornography, Technology, and Progress," *ICON* 4 (1998).

7. Mare Wielage and Rod Woodcock (authors of "The Rise and Fall of Beta"), interview with the author, March 15, 2003.

8. The canonical reference on the VHS/Beta format war is M. Cusumano, Y. Mylonadis, and R. Rosenbloom, "Strategic Maneuvering and Mass-Market Dynamics—The Triumph of VHS over Beta," *Business History Review* 66, no. 1 (1992).

9. Jim Lowe, The Videophile's Newsletter, December 1976, 3.

10. Mike Salomon Video retreat, interview with the author, May 10, 2003.

11. "How to Care for Your Videocassettes," *Video Magazine*, Summer 1978.

12. Lardner, *Fast Forward*, 96.

13. "Blank Tape: The Current Market," *Video Magazine*, Summer 1978.

14. "Fuji Videocassette Advertisement," *Video*, Summer 1979.

15. The distinction seems to have grown out of an older definition of "white goods" as domestic linens, which reaches at least as far back as the beginning of the twentieth century. The Oxford English Dictionary offers an example of the term's usage to describe "refrigerators, deep freezers, washing machines, [and] clothes dryers" in 1960, as well as a reference from 1976 that drew a distinction between such white goods and "TVs and audio (which the trade calls brown goods)." See "white, a."*Oxford English Dictionary Online*. Oxford University Press, http://dictionary.oed.com.

16. Cynthia Cockburn and Susan Ormrod, *Gender and Technology in the Making* (London: Sage Publications, 1993), esp. chapter 4.

17. On the history of home video games in the 1970s and 1980s, see Scott Cohen, *Zap! The Rise and Fall of Atari* (New York: McGraw-Hill Book Company, 1984); J.C. Herz, *Joystick Nation: How Videogames Ate Our Quarters, Won Our Hearts, and Rewired Our Minds* (New York: Little Brown & Company, 1997); Steven Poole, *Trigger Happy: Videogames and the Entertainment Revolution* (New York: Arcade Books, 2000); David Sheff, *Game Over: How Nintendo Zapped an American Industry, Captured Your Dollars, and Enslaved Your Children* (New York: Random House, 1993); Mark J. P. Wolf, ed., *The Medium of the Video Game* (Austin, Texas: University of Texas Press, 2001).

18. "Blank Tape: The Current Market."

19. The distinction between a distributor and a manufacturer's "rep" boiled down to where the orders were sent once placed—reps would take orders from retailers on behalf of specific manufacturers and pass them on to manufacturers or wholesalers for a commission, while distributors would buy goods from manufacturers themselves, maintaining a warehouse from which they would fill their own orders. As one rep described himself, "The rep is the arms and legs of the manufacturer . . . when you are dealing with the rep, you *are* dealing directly with the manufacturer." Distributors, on the other hand, were autonomous agents, buying from manufacturers whatever goods they chose to carry, and selling them to retailers on their own

terms ("Dick Turchen: A Talk with a 'Super Rep'," *The Video Programs Retailer*, Fall 1980).

20. Interestingly, Peter Keane had left Sony in 1971 and within a year was working for Cartrivision, where he brought his expertise to both the technical design and the marketing of the Cartrivision player. Additionally, he leveraged his contacts at Columbia Studios to broker the licensing deal that allowed Cartrivision to distribute tapes of many Columbia movies. (Keane, interview.)

21. Lardner, *Fast Forward*, 83–84.

22. Keane, interview.

23. Nmungwun, *Video Recording Technology*, page #. Chapter 6.

24. Lardner, *Fast Forward*, 85.

25. Keane, interview.

26. Graham, *RCA and the Videodisc*, 142–143.

27. "News and Comment," *Videography*, September 1976.

28. Noel Gimbel (owner of Sound Unlimited), interview with the author, March 12, 2003.

29. Jeff Tuckman (film collector and Sound Unlimited employee), interview with the author, December 9, 2003.

30. Eric Schlosser, *Reefer Madness* (Boston: Houghton Mifflin, 2003), 126–128.

31. Frederick S. Lane III, *Obscene Profits: The Entrepreneurs of Pornography in the Cyber Age* (New York: Routledge, 2000).

32. Samuel R. Delany, *Times Square Red, Times Square Blue* (New York: New York University Press, 1999).

33. On the general subject of pornography in various media, see James Elias, Veronica Diehl Elias, Vern L. Bullough, Gwen Brewer, Jeffrey J. Douglas, and Will Jarvis, eds., *Porn 101: Eroticism, Pornography, and the First Amendment* (New York: Prometheus Books, 1999); David Hebditch and Nick Anning, *Porn Gold: Inside the Pornography Business* (London: Faber and Faber, 1988); Walter Kendrick, *The Secret Museum: Pornography in Modern Culture* (New York: Viking, 1987); Laurence O'Toole, *Pornocopia: Porn, Sex, Technology and Desire* (London: Serpent's Tail, 1998).

34. Schlosser, *Reefer Madness*, 138.

35. Howard Polskin, "Tuning into the Videotape Scene," *Playboy*, April 1979.

36. Tuckman, interview.

37. Ibid.

38. Neal Rosen, "A Tribute to the Founding Fathers: Andre Blay," *Video Software Dealer*, July 1986.

39. Lardner, *Fast Forward*, 172–173.

40. Rosen, "A Tribute to the Founding Fathers: Andre Blay."

41. "Video Club of America Advertisement," *TV Guide*, November 26, 1977.

42. Ellipses in original text.

43. "The Club," *Videography*, November 1977.

44. Rosen, "A Tribute to the Founding Fathers: Andre Blay."

45. Joel A. Samberg, "Distributor Interview: Video Program Distribution—The Busiest Business on the Block," *The Video Programs Retailer*, January 1981.

46. Wayne Mogel (B&S Sales' only salesman), interview with the author, April 30, 2003.

47. This term was quickly picked up by specialty video magazines; for example, the Winter 1978 *Video Magazine Buyer's Guide* (the first issue of the magazine, which appeared in early 1978) contained the words "Video Software" splashed across two pages, marking the section profiling prerecorded videocassettes.

48. Mogel, interview.

49. Lardner, *Fast Forward*, 174–175.

50. Rosen, "A Tribute to the Founding Fathers: Andre Blay."

51. Frederick Wasser, *Veni, Vidi, Video* (Austin: University of Texas Press, 2001), 96.

52. Mogel, interview.

53. Don Rosenberg (Schwartz Brothers Salesman), interview with the author, March 18, 2003.

54. Ibid.

55. Ibid.

56. It's worth noting that the book industry operated on the same principle as the music industry, also treating its goods as texts rather than artifacts.

57. Jerry Frebowitz (Movies Unlimited Founder and President), interview with the author, May 28, 2003.

58. Rosenberg, interview.

59. Ibid.

60. Ibid.

61. On home economists and sales agents, see Carolyn M. Goldstein, "From Service to Sales: Home Economics in Light and Power, 1920–1940," *Technology and Culture* 38, no. 1 (1997); Kline, *Consumers in the Country: Technology and Social Change in Rural America* Ronald R. (Baltimore: Johns Hopkins University Press 2000); and on rural sales agents, see Zunz, *Making America Corporate, 1870–1920* (Chicago: University of Chicago Press, 1990).

62. Trevor Pinch and Frank Trocco, *Analog Days: The Invention and Impact of the Moog Synthesizer* (Cambridge, MA: Harvard University Press, 2002), chapter 11: "Inventing the Market."

63. Ibid, 244.

64. Ibid, 247.

65. Wasser, *Veni, Vidi, Video*, 96.

66. Bruno Latour, "Give Me a Laboratory and I Will Raise the World," in *Science Observed: Perspectives on the Social Study of Science*, ed. Karin D. Knorr-Cetina and Michael Mulkay (Beverly Hills: Sage, 1983); see also Michel Callon, "Some Elements of a Sociology of Translation: Domestication of the Scallops and the Fishermen of St Brieuc Bay," in *Power, Action and Belief: A New Sociology of Knowledge?*, ed. John Law (London: Routledge & Kegan Paul, 1986).

Chapter 3

1. Ruth Schwartz Cowan, "The Consumption Junction: A Proposal for Research Strategies in the Sociology of Technology," in The Social Construction of Technological Systems, ed. Trevor Pinch, Wiebe Bijker, and Thomas Hughes (Cambridge, MA: MIT Press, 1987).

2. Others who've used the "consumption junction" as an analytic tool include Carolyn M. Goldstein, "From Service to Sales: Home Economics in Light and Power, 1920–1940," *Technology and Culture* 38, no. 1 (1997); and Karin Zachmann, "A Socialist Consumption Junction: Debating the Mechanization of Housework in East Germany, 1956–1957," *Technology and Culture* 43, no. 1 (2002).

3. "Selling Software in the Big Apple," *Video Store*, January 1980.

4. Joel A Samberg, "Retailer Case Study: The Video Shack," *The Video Programs Retailer*, January 1981.

5. Lance Strate (professor of communication and media studies, Fordham University), interview with the author, February 27, 2003. For an extensive analysis of salespeople and the "hard sell," see Trevor Pinch and Colin Clark, *The Hard Sell: The Language and Lessons of Street-Wise Marketing* (London: HarperCollins, 1988).

6. Samberg, "Retailer Case Study: The Video Shack."

7. Lardner, *Fast Forward*, 176.

8. Ibid, 175–178; David Rowe, "The Real George Atkinson," *Video Store*, June 1983.

9. "Fotomat to Carry Programs," *Videography*, September 1978.

10. "Fotomat Launches Drive-Thru Movies Nationally," *Videography*, December 1979.

11. *United States v. Paramount Pictures, Inc.*, Nos. 79 to 86 (1948).

12. "Fotomat Launches Drive-Thru Movies Nationally."

13. Jerry Frebowitz (Movies Unlimited founder and president), interview with the author, May 28, 2003.

14. Noel Gimbel (owner of Sound/Video Unlimited), interview with the author, March 12, 2003.

15. Michael Becker (owner of The Sound Room), interview with the author, April 23, 2003.

16. Tom Adams, "What's So Hot About Erol's," *Video Store*, December 1984.

17. "Regional Reports," *Video Store*, September 1980.

18. See "Video Store Top 12: Risley's Audio & Video," *Video Store*, January 1984.

19. Becker, interview.

20. "Video Dynamics and the Tricks of the Trade," *The Video Programs Retailer*, December 1981.

21. "American Marketplace: Spec's Dixie Highway," *Video Store*, October 1985; "Salzer's Video," *Video Store*, December 1985.

22. "Regional Reports," *Video Store*, April 1983.

23. "Jack Messer's Video Store: Following His Own Footsteps," *The Video Programs Retailer*, September 1981; "Regional Reports," *Video Store*, September 1984.

24. Henry Tolino (owner of The Media Center), interview with the author, May 15, 2003.

25. "Home Video in the Bag," *Video Business*, April 1984.

26. Mark Fisher and Carrie Dieterich (of the Video Software Dealers Association), interview with the author, March 10, 2003.

27. "Home Video in the Bag."

28. Ibid.

29. Lisa Lilienthal, "Hooray for Haullywood," *Video Store*, 1984.

30. David Linck, "The Home Video Explosion—Will Exhibitors Let It Pass By?" *Boxoffice*, December 1981.

31. Alan Karp, "Selling Video in the Theatre," *Boxoffice*, June 1984, 19.

32. See, for example, Randy Lewis, "Movies: Popcorn and Videos," *Los Angeles Times*, August 30, 1982; "The Silver Screen and the Video Monitor," *Video Store*, December 1984.

33. Karp, "Selling Video in the Theatre," 9.

34. "Theater Chain, Franchiser in 67-Store Deal," *Video Business*, August 1982.

35. Ron Berger (National Video), Interview with the author, March 21, 2003.

36. Robert Klingensmith, "Selling Videotapes Isn't Selling Out," *Box Office*, May 1984.

37. "Advertisement," *Videography*, February 1979.

38. Frederick Wasser, *Veni, Vidi, Video* (Austin: University of Texas Press, 2001) 68.

39. Ron Berger, interview.

40. Lardner, *Fast Forward, 185.*

41. Video Consultants, The 1980 National Video Survey.

42. "The Video Shoppe," *Video Store*, January 1980.

43. Joel A Samberg, "Retailer Case Study: That's Entertainment," *The Video Programs Retailer*, March 1981.

44. Mike Salomon (camera and video store owner), interview with the author, May 10, 2003.

45. Steve Savage (founder and owner of New Video), interview with the author, March 3, 2003.

46. Brad Burnside (founder, Video Adventures), interview with the author, June 4, 2003.

47. Mitch Lowe (owner and founder, Video Droid), interview with the author, April 3, 2003.

48. Daniel Loren Moret, "The New Nickelodeons: A Political Economy of the Home Video Industry with Particular Emphasis on Video Software Retailers" (masters thesis, University of Oregon, 1991), 35.

49. Douglas Gomery, *Shared Pleasures: A History of Movie Presentation in the United States* (Madison: University of Wisconsin Press, 1992), 31.

50. Megumi Komiya and Barry Litman, "The Economics of the Prerecorded Videocassette Industry," in *Social and Cultural Aspects of VCR Use*, ed. Julia R. Dobrow (Hillsdale, NJ: Lawrence Erlbaum Associates, 1990), 34.

51. Tolino, interview.

52. Rosenberg (Schwartz Brothers Salesman), interview with the author, March 18, 2003.

53. Wayne Mogel (of Star Video), interview with the author, April 30, 2003.

54. This resonates with Fred Turner's description of Stewart Brand as a "network entrepreneur"—see Fred Turner, *From Counterculture to Cyberculture: Stewart Brand, the Whole Earth Network, and the Rise of Digital Utopianism* (University of Chicago Press, 2006).

55. Mogel, interview.

56. That's not to say that this advice was universal—as Mogel himself recalls, "You had different demographics . . . what would work up in Spanish Harlem wasn't necessarily going to work in Midtown Manhattan." (Ibid).

57. Becker, interview.

58. "Homer H. Hewitt, III: The Expert's Expert," *The Video Programs Retailer*, November 1981.

59. Mogel, interview.

60. Ibid.

Chapter 4

1. On the social and cultural history of the movie theater as an American institution, see Douglas Gomery, *Shared Pleasures: A History of Movie Presentation in the United States* (Madison: University of Wisconsin Press, 1992) and Robert Sklar, *Movie-Made America: A Cultural History of American Movies* (New York: Vintage, 1994).

2. "Putting Theaters in Stores," *Video Store*, November 1985.

3. David Crook, "'Let's Take in a (Cassette) Show'," *Los Angeles Times*, September 7, 1982.

4. D.A.S., "The Theater Experience," *Video Store*, April 1984.

5. Herb Fischer, "Speaking Out," *Video Software Dealer*, March 1987.

6. For more on the history of popcorn in movie theaters, see Gomery, *Shared Pleasures*, chapter 5; Andrew F. Smith, *Popped Culture: A Social History of Popcorn in America* (Columbia, SC: University of South Carolina Press, 1999).

7. Smith, *Popped Culture*, chapter 7.

8. "Advertisement for Orville Redenbacher Popcorn," *Video Business* 1987.

9. Eve Allsbrook, "The Video Convenience Store Is Here!" *Video Software Dealer*, March 1987, 39; "Regional Reports," *Video Store*, December 1982.

10. Allsbrook, "The Video Convenience Store Is Here!," 35.

11. Jill Gottesman, "Time for a Snack," *Video Store*, April 1987, 60.

12. Ken Dorrance, (of the Video Software Dealers Association), interview with the author, April 3, 2003; "Pizza Made Behind Tapes," *Video Store*, September 1987.

13. "Rental vs. Sale: The Controversy Grows in the Software Market," *Video Store*, May 1980.

14. *VCRs and the American Way-of-Life* (Glendale, CA: The Barna Research Group, 1988).

15. "End of 'Lifetime Rentals' as Star Wars Comes up for Sale," *Video Business*, August 1982.

16. "Our View: The Selling of Star Trek II," *Video Business*, January 1983.

17. "*Flashdance*: What a Feeling to Make a Sale," *Home Entertainment Marketing*, September 1983.

18. Frederick Wasser, *Veni, Vidi, Video* (Austin: University of Texas Press, 2001), 134.

19. Rich Nathanson (clerk, New Video) interview with the author, March 13, 2003.

20. Ibid.

21. Laurie Garcia (owner, Midtown Video) interview with the author, March 31, 2003. Brandon Hart, online survey filed at The Video Store Project website, March 6, 2003, available at http://echo.gmu.edu/workshops/jgreenberg/viewdetail.php?responseID=297; Nick MacPherson, online survey filed at The Video Store Project website, February 16, 2003, available at http://echo.gmu.edu/workshops/jgreenberg/viewdetail.php?responseID=74; Damon Plummer, online survey filed at The Video Store Project website, February 12, 2003, available at http://echo.gmu.edu/workshops/jgreenberg/viewdetail.php?responseID=46.

22. Alex Blair, online survey filed at The Video Store Project website, March 6, 2003, available at http://echo.gmu.edu/workshops/jgreenberg/viewdetail.php?responseID=474; Geoffrey Brown, online survey filed at The Video Store Project website, March 6, 2003, available at http://echo.gmu.edu/workshops/jgreenberg/viewdetail.php?responseID=211; Lucy Dunne, online survey filed at The Video Store Project website, March 7, 2003, available at http://echo.gmu.edu/workshops/jgreenberg/viewdetail.php?responseID=666; Jameson Wetmore, online survey filed at The Video Store

Project website, February 10, 2003, available at http://echo.gmu.edu/workshops/ jgreenberg/viewdetail.php?responseID=14.

23. John Martin, online survey filed at The Video Store Project website, March 7, 2003, available at http://echo.gmu.edu/workshops/jgreenberg/viewdetail.php?responseID= 738.

24. Armory was the first manufacturer of the ubiquitous plastic boxes (also known as "clamshells") that opened from one side to hold a single videocassette, and even though countless other manufacturers eventually joined them, "Armory case" became a catch-all retailer term for such cases (Michael Becker (owner of the Sound Room), interview with the author, April 23, 2003).

25. Ed Hulse, "The VCR Rental Revival," *Video Business*, October 1983.

26. David Allen Shaw, "The 93 Percent Solution," *Video Store*, July 1983.

27. D. R. Martin, "Big Ticket Rentals," *Video Store*, July 1986.

28. "Industry News: Lower Prices for Rentabeta," *Video Store*, July 1982.

29. Shaw, "The 93 Percent Solution."

30. It's not clear, however, exactly how a customer who recorded a "special show" would *watch* said tape once the RentaBeta or PortaVideo unit was returned to the video store.

31. Ryan Beeman, online survey filed at The Video Store Project website, March 6, 2003, available at http://echo.gmu.edu/workshops/jgreenberg/viewdetail.php? responseID=325; Kevin Boury, online survey filed at The Video Store Project website, March 6, 2003, available at http://echo.gmu.edu/workshops/jgreenberg/viewdetail .php?responseID=399; Daniel Kern, online survey filed at The Video Store Project website, March 6, 2003, available at http://echo.gmu.edu/workshops/jgreenberg/ viewdetail.php?responseID=182; Mike McLean, online survey filed at The Video Store Project website, March 6, 2003, available at http://echo.gmu.edu/workshops/jgreen-berg/viewdetail.php?responseID=284; Ty Nelson, online survey filed at The Video Store Project website, March 6, 2003, available at http://echo.gmu.edu/workshops/ jgreenberg/viewdetail.php?responseID=366.

32. See chapter 2's discussion of the stag film and film technology.

33. On the role of the television in twentieth century American homes, see Matthew Geller and Reese Williams, eds., *From Receiver to Remote Control: The TV Set* (New York: The New Museum of Contemporary Art, 1990); Lynn Spigel, *Make Room for TV: Television and the Family Ideal in Postwar America* (Chicago: University of Chicago Press, 1992); Cecelia Tichi, *Electronic Hearth: Creating an American Television Culture* (New York: Oxford University Press, 1991). On the introduction of the VCR into this domestic space, see Amy B. Jordan, "A Family Systems Approach to the Use of the

VCR," in *Social and Cultural Aspects of VCR Use*, ed. Julia R. Dobrow (Hillsdale, NJ: Lawrence Erlbaum Associates, 1990).

34. Spigel, *Make Room for TV*, 38–39.

35. Esther B. Fein, "Of Popcorn, Loneliness, and VCRs," *The New York Times*, August 27, 1985.

36. Ibid.

37. For an empirical (and quantitative) study of the actual practices of home viewers, see Milton J. Shatzer and Thomas R. Lindlof, "Subjective Differences in the Use and Evaluation of the VCR," in *The VCR Age*, ed. Mark R. Levy (Newbury Park: SAGE Publications, 1989).

38. Fein, "Of Popcorn, Loneliness, and VCRs."

39. Gene Siskel, "Movie Manners," *Ladies Home Journal*, May 1982.

40. Richard Sandomir, "A Tragedy of Manners at the Movies," *The New York Times*, June 5, 1992.

41. On the implications of VCR use for the power dynamics of the parent-child relationship, see Katharine E. Heintz, "VCR Libraries: Opportunities for Parental Control," in *Social and Cultural Aspects of VCR Use*, ed. Julia R. Dobrow (Hillsdale, NJ: Lawrence Erlbaum Associates, 1990).

42. Rick Anguilla, "Kidvid: It's Serious Business," *Video Retailing*, April 1982.

43. *Selectavision Promotional Still* (Sarnoff Archives, Princeton).

44. Dobrow offers an analysis of the tendency for VCR owners to watch particular cassettes multiple times (Julia R. Dobrow, "The Rerun Ritual: Using VCRs to Re-View," in *Social and Cultural Aspects of VCR Use*, ed. Julia R. Dobrow [Hillsdale, NJ: Lawrence Erlbaum Associates, 1990]). On children's video in particular, see Christopher Paul Denis, "Spinnaker Swings into Educational Video," *Video Business*, December 1985; Ed Hulse, "Children's Video: A Market Comes of Age," *Video Business*, December 1985.

45. Steve Chagollan, "The Half-Inch Babysitter," *Video Store*, November 1985.

46. "Adult Software Roundtable," *Video Retailing*, March 1982; Machael Hoff, "Adult Video: Solid as a Rock," *Video Business*, June 1985; Megumi Komiya and Barry Litman, "The Economics of the Prerecorded Videocassette Industry," in *Social and Cultural Aspects of VCR Use*, ed. Julia R. Dobrow (Hillsdale: Lawrence Erlbaum Associates, 1990, 36).

47. Michael Horenstein, "Video Centerfolds," *The Video Retailer*, August/September 1982; "Playmates without Staples," *The Video Retailer*, December 1982.

48. "Conversation with Al Goldstein," *Videography*, January 1979.

49. Michael deCourcy Hinds, "Starring in Tonight's Erotic Video: The Couple Down the Street," *The New York Times*, March 22, 1991.

Chapter 5

1. Ray Oldenburg, *The Great Good Place: Cafés, Coffee Shops, Community Centers, Beauty Parlors, General Stores, Bars, Hangouts, and How They Get You through the Day*, 2nd ed. (New York: Marlowe & Co., 1997), 16.

2. Ibid, xvii.

3. Ibid, 26.

4. Ibid, 40.

5. Richard Butsch, *The Making of American Audiences: From Stage to Television, 1750–1990* (New York: Cambridge University Press, 2000), 12–13.

6. Ibid, 134.

7. For a general look at the role of theaters in American cultural life, see Ibid; Douglas Gomery, *Shared Pleasures: A History of Movie Presentation in the United States* (Madison: University of Wisconsin Press, 1992).

8. On nickelodeons, see Butsch, *The Making of American Audiences: From Stage to Television, 1750–1990*, chapter 10, "The Celluloid Stage: Nickelodeon Audiences"; Gomery, *Shared Pleasures*, chapter 2, "The Nickelodeon Era."

9. For more on the role of motion pictures in small town America, see Kathryn H. Fuller, *At the Picture Show: Small-Town Audiences and the Creation of Movie Fan Culture* (Washington, DC: Smithsonian Institution Press, 1996).

10. Butsch, *The Making of American Audiences*, 172.

11. On drive-in theaters, see Kerry Segrave, *Drive-in Theaters:A History from Their Inception in 1933* (Jefferson, NC: McFarland & Company, 1992).

12. See for example the description of *Pop*Card* in "Regional Reports," *Video Store*, January 1985.

13. Barbara Dianne Savage, *Broadcasting Freedom: Radio, War, and the Politics of Race, 1938–1948* (Chapel Hill: University of North Carolina Press, 1999).

14. Laurie Garcia (owner, Midtown Video) interview with the author, March 31, 2003.

15. Steve Savage (founder and owner of New Video), interview with the author, March 3, 2003.

16. Stencel (former video clerk) interview with the author, January 31, 2003.

17. Jacques Ellul, *Propaganda: The Formation of Men's Attitudes* (New York: Vintage Books, 1973).

18. For more on the relationship between membership fees and customer attachment to a given store, see Alan S. Dick, "Using Membership Fees to Increase Customer Loyalty," *The Journal of Product and Brand Management* 4, no. 5 (1995); Alan S. Dick and Kenneth R. Lord, "The Impact of Membership Fees on Consumer Attitude and Choice," *Psychology & Marketing* 15, no. 1 (1998).

19. Michael Miller, "How to Produce a First-Class Newsletter," *Video Store*, June 1984.

20. Dean Stevens, "All the News That Fits," *Video Store*, September 1986.

21. "Regional Reports."

22. Michael Becker (owner of The Sound Room), interview with the author, April 23, 2003.

23. Charles Wesley Orton, "Tricks from the Pizza Trade," *Video Store*, September 1987.

24. Mimi Kmet, "Video to Go! Service on Wheels," *Video Store Dealer*, February 1986; Orton, "Tricks from the Pizza Trade," 74–81; "Storefronts: Express Video," *Video Store*, August 1986.

25. Michael Dark (video entrepreneur), interview with the author, March 31, 2003.

26. Garcia, interview.

27. Ken Dorrance (of the Video Software Dealers Association and owner, Video Station), interview with the author, April 3, 2003.

28. Jack Stein and Bernice Stein (video store owners in Huntington Beach) interview with the author, March 16, 2003.

29. Dark, interview.

30. Steve Savage, interview.

31. Dark, interview.

32. Dorrance, interview.

33. Steve Savage, interview.

34. Mitch Lowe (founder and owner, Video Droid) interview with the author, April 3, 2003.

35. Becker, interview.

36. "Storefronts: Express Video." (of Video Adventures), interview with the author, April 2, 2003.

37. Becker, interview; Brad Burnside (of Video Adventures), Adrian Hickman interview with the author, May 2003.

38. Chris Ritter (of Video Droid), interview with the author, April 3, 2003.

39. Stencel, interview.

40. Ritter, interview.

41. Jerry Frebowitz (and president, Movies Unlimited founder), interview with the author, May 28, 2003.

42. Dorrance, interview.

43. Matt Ratto (son of Gary Ratto), interview with the author, March 17, 2003.

44. Stencel, interview.

45. Mimi Kmet, "Becoming America's Gourmet Video Store," *Video Software Dealer*, September 1985; Rich Nathanson (of New Video), interview with the author, March 13, 2003. Savage, interview.

46. Paul Fishbein (clerk at Movies Unlimited, and Farder, Adult Video News) interview with the author, April 15, 2003.

47. Dark, interview.

48. Ray Pennisi, online survey filed at The Video Store Project website, March 13, 2003, available at http://echo.gmu.edu/workshops/jgreenberg/viewdetail.php?responseID=907.

49. Frebowitz, interview.

50. This in-store environment was strikingly similar to the world of the specialty record store, perhaps best described in Nick Hornby's novel *High Fidelity* (New York: Riverhead Books, 1995).

51. Steve Savage, interview.

52. Dorrance, interview.

53. Becker, interview.

54. Ritter, interview.

55. Ibid.

56. For a detailed look at this period in Tarantino's life, see Jami Bernard, *Quentin Tarantino: The Man and His Movies* (New York: Harper Perennial, 1995), chapter 3.

57. Jay Carr, "Tarantino's New Cool '90s," *The Boston Globe*, October 9, 1994.

58. Bernard, *Quentin Tarantino*, 30–31.

59. Bob Flynn, "The Counter Culture That Made It at Cannes," *Sunday Times* (London), August 14, 1994.

60. Fishbein, interview.

61. Oldenburg, *The Great Good Place*, 170.

62. Ratto, interview.

63. Ibid.

64. Lowe, interview.

65. Stencel, interview.

66. Stephen Bojanowski, online survey filed at The Video Store Project website, March 10, 2003, available at http://echo.gmu.edu/workshops/jgreenberg/viewdetail .php?responseID=867.

67. Charles Daevlyn, online survey filed at The Video Store Project website, March 6, 2003, available at http://echo.gmu.edu/workshops/jgreenberg/viewdetail.php? responseID=314.

68. unknown, online survey filed at The Video Store Project website, March 6, 2003, available at http://echo.gmu.edu/workshops/jgreenberg/viewdetail.php?responseID= 272.

69. Nathanson, interview.

70. Pat Nestor, online survey filed at The Video Store Project website, March 6, 2003, available at http://echo.gmu.edu/workshops/jgreenberg/viewdetail.php?responseID= 591.

71. Savage, interview.

72. Dark, interview.

73. Lance Strate (professor of communication and media studies, Fordham University), interview with the author, February 27, 2003.

74. Frebowitz, interview.

75. Shawn Bowles, online survey filed at The Video Store Project website, March 6, 2003, available at http://echo.gmu.edu/workshops/jgreenberg/viewdetail.php? responseID=351.

76. Stencel, interview.

77. Nathanson, interview.

78. Savage, interview.

79. Steve Savage, interview.

80. "Regional Reports." *Video Store*, May 1984, 64–65.

81. "Regional Reports." *Video Store*, September 1984, 110.

82. Ritter, interview.

Chapter 6

1. See, among others: Wiebe E. Bijker, *Of Bicycles, Bakelites, and Bulbs: Toward a Theory of Sociotechnical Change* (Cambridge, MA: MIT Press, 1995); Ronald Kline and Trevor Pinch, "Users as Agents of Technological Change: The Social Construction of the Automobile in the Rural United States", *Technology and Culture* 37, no. 4 (October 1996); Wiebe E. Bijker, Thomas Parke Hughes, and T. J. Pinch, *The Social Construction of Technological Systems: New Directions in the Sociology and History of Technology* (Cambridge, MA: MIT Press, 1987); Wiebe E. Bijker and John Law, *Shaping Technology/ Building Society: Studies in Sociotechnical Change* (Cambridge, MA: MIT Press, 1992).

2. Ruth Schwartz Cowan, "The Consumption Junction: A Proposal for Research Strategies in the Sociology of Technology," in *The Social Construction of Technological Systems*, eds. Wiebe E. Bijker, Thomas Park Hughes, and T. J. Pinch (Cambridge, MA: MIT Press, 1987).

3. *Time* magazine, for instance, included a special home video supplement with its October 30, 1978 issue.

4. Frank Barnako (founder, Potomac Video), interview with the author, April 26, 2003.

5. Michael Dark (Video enterprener), interview with the author, March 31, 2003.

6. Michael Becker (founder, The Sound Room), interview with the author, April 23, 2003.

7. Ken Winslow, "Winter CES: A Program Retailer's Perspective," *The Video Programs Retailer*, March 1981.

8. Seth Goldstein, "A Gamble or a Sure Bet?" *Video Business*, December 1985.

9. Noel Gimbel (owner, Sound/Video Unlimited), interview with the author, March 12, 2003.

10. Emil Knodell, "The Mathias Amendment: Last Rites for First Sale," *Video Store*, May 1982.

11. "MGM/CBS Launches First Run Home Video Theater," *Video Store*, February 1982; "Warner Goes Rental-Only," *Video Store*, November 1981.

12. "Choosing up Sides," *Video Store*, June 1983; Knodell, "The Mathias Amendment."

13. "VSDA Welcomes Regional Group," *Video Store*, January 1983.

14. "Warner Goes Rental-Only."

15. James Lardner, *Fast Forward: Hollywood, the Japanese, and the Onslaught of the VCR* (New York: Norton, 1987), 197

16. Bob Davis, "Texas Retailers 'Boycott' Warner," *Dealerscope II*, November 1981.

17. Lardner, *Fast Forward*, 194

18. Ibid, 199

19. "Distributor/Retailer Updates," *The Video Retailer*, December 1982.

20. "NARM 1981 Video Retailers Convention," *The Video Programs Retailer*, September 1981.

21. "Special Report: NARM Video Convention," *Video Retailing*, September 1981.

22. Gimbel, interview; Jeff Tuckman (film collector and employee, Sound Unlimited), interview with the author, December 9, 2003.

23. "Software Spotlight: The Video Software Retailers Association," *Video Store*, February 1983.

24. "VSDA Welcomes Regional Group."

25. Don Rosenberg (Schwartz Brothers Salesman), interview with the author, March 18, 2003.

26. "Video Reporter: No Soap in Dallas . . . " *Video Store*, September 1982.

27. Goldstein, "A Gamble or a Sure Bet?"

28. Mike Salomon (camera and video store owner), interview with the author, May 10, 2003.

29. "Video Dynamics and the Tricks of the Trade."

30. Dark, interview.

31. Ibid.

32. David Rowe, "The Real George Atkinson," *Video Store*, June 1983.

33. Ken Dorrance (of the Video Software Dealers Association and owner, Video Station Oak land, CA), interview with the author, April 3, 2003.

34. "Regional Reports."

35. This section relies heavily on an interview I conducted with Ron Berger (head of National Video), March 2, 2003.

36. Among his customers were Hollywood studios, major television networks like ABC and CBS, and perhaps most unlikely, television parts manufacturers including Westinghouse and Emerson. As Berger explains, "The TV bulb guys were trying to figure out, you know, 'If somebody plays these videos on our TVs all day long, what would that do?'" (Berger, interview).

37. "Regional Reports."

38. Berger, interview.

39. Eric Schlosser, *Fast Food Nation* (New York: Harper Collins, 2002), chapter 1.

40. Thomas P. Hughes, "Technological Momentum," in *Does Technology Drive History?* ed. Merritt Roe Smith and Leo Marx (Cambridge, MA: MIT Press, 1994).

41. This section relies mainly on Gail DeGeorge, *The Making of a Blockbuster* (New York: Wiley, 1995 particularly chapters 5 and 6.

42. Ironically, video games were rented in Blockbuster stores, and nobody ever seemed to question how exactly they fit into the general aesthetic. By this point, video games had become such a routine presence in video stores that their non-Hollywood orientation was simply overlooked, if not ignored altogether.

43. Rich Nathanson (of New Video), interview with the author, March 13, 2003.

44. Ibid.

45. Mark Wattles (founder of Hollywood Video), interview with the author, April 29, 2003.

46. Even these independents embraced the franchise ideal, though on a much smaller scale: "In 1982, the typical retailer owned one outlet. The 1988 VSDA study pegs the average 'independent' with 6.8 stores." (David Rowe, "Speaking Out," *Video Software Dealer*, May 1988.)

Chapter 7

1. Frederick Wasser, *Veni Vidi Video* (Austin: University of Texas Press, 2001).

2. The roots of this primacy extended far deeper than mere economics, and the rhetoric of theater versus ancillary media persisted long after the theater was no longer the dominant source of revenue.

3. At the same time as the establishment of video stores, cable and satellite networks offered another distinct option to viewers, but as these subscription-based services

were essentially constructed as "television plus," I can lump them in with television for the purposes of my argument here.

4. McLuhan, *Understanding Media: The Extensions of Man*, 1st MIT Press ed. (Cambridge, MA: MIT Press, 1994).

5. J. David Bolter and Richard Grusin, *Remediation: Understanding New Media* (Cambridge, MA: MIT Press, 1999).

6. Ibid.

7. Carolyn A. Lin, "Audience Activity and VCR Use," Kimberly K. Massey and Stanley J. Baran, "VCRs and People's Control of Their Leisure Time," both in *Social and Cultural Aspects of VCR Use*, ed. Julia R. Dobrow (Hillsdale, NJ: Lawrence Erlbaum Associates, 1990).

8. For an overview, see Neil Postman, *Amusing Ourselves to Death* (New York: Penguin Books, 1985).

9. From here onward, I use the capitalized version to refer to this idealized form of a movie, as opposed to individual manifestations of a given movie in practice.

10. Michele Hilmes, *Hollywood and Broadcasting: From Radio to Cable* (Urbana: University of Illinois Press, 1990); William Lafferty, "Feature Films on Prime-Time Television," in *Hollywood in the Age of Television*, ed. Tino Balio (Boston: Unwin Hyman, 1990).

11. Judith Crist, "Movies," *TV Guide*, September 10, 1977.

12. Janet Maslin, "Critic's Notebook: Cable TV, Home Video and Chopping of Movies," *New York Times*, March 25, 1982.

13. With the exception, of course, of the made-for-TV movie, a genre unto itself that often relied on the frequent commercial interruptions for its narrative structure; see Laurie Schulze, "The Made-for-TV-Movie: Industrial Practice, Cultural Form, Popular Reception," in *Hollywood in the Age of Television*, ed. Tino Balio (Boston: Unwin Hyman, 1990).

14. "AFI Report," (Los Angeles: American Film Institute, 1971).

15. On the early history of advertiser-supported broadcasting, see Susan Smuylan's *Selling Radio: The Commercialization of American Broadcasting, 1920–1934* (Washington, DC: Smithsonian Institution Press, 1994).

16. "Commercials on Cassette: Consumers Speak Out," *Video Business*, January 1985.

17. For a more extensive discussion of VCR users' relationship with their fast-forward buttons, see Bruce C. Klopfenstein, "Audience Measurement in the VCR

Environment: An Examination of Ratings Methodologies," in *Social and Cultural Aspects of VCR Use*, ed. Julia R. Dobrow (Hillsdale, NJ: Lawrence Erlbaum Associates, 1990); Lin, "Audience Activity and VCR Use"; Barry S. Sapolsky and Edward Forrest, "Measuring VCR 'Ad-Voidance'," in *The VCR Age: Home Video and Mass Communication*, ed. Mark R. Levy (Newbury Park: Sage Publications, 1989).

18. "Scanning the Regions: How Do You Feel About Commercials on Videocassettes?" *Video Software Dealer*, March 1987.

19. "Commercials on Cassette: Consumers Speak Out."

20. Wasser, *Veni, Vidi, Video*, 145.

21. "Scanning the Regions: How Do You Feel About Commercials on Videocassettes?"

22. See John Belton, *Widescreen Cinema* (Cambridge, MA: Harvard University Press, 1992).

23. John Belton, "Glorious Technicolor, Breathtaking Cinemascope, and Stereophonic Sound," in *Hollywood in the Age of Television*, ed. Tino Balio (Boston: Unwin Hyman, 1990); Belton, *Widescreen Cinema*, 44–45.

24. See, for example, R. M. Hayes, *3-D Movies* (Jefferson, NC: McFarland & Company Inc., 1989). In addition, several essays on even more contemporary theatrical technologies can be found in Philip Hayward and Tana Wollen, eds., *Future Visions: New Technologies of the Screen* (London: British Film Institute, 1993).

25. Marc Wielage and Rod Woodcock (authors of "The Rise and Fall of Beta"), interview with the author, March 15, 2003.

26. "Video Club of America Advertisement;" *TV Guide*, November 26, 1977.

27. Lowell Goldman, "Directors on Video," *BoxOffice*, March 1985.

28. Woody Allen (via Allen Eichhorn, publicist/agent), personal correspondence (2004).

29. Wielage and Woodcock, interview.

30. The gray bars (as opposed to black) are an interesting story of boundary drawing between medium and message in their own right: according to Wielage, "[A United Artists studio executive] said, 'You know, I don't think we should put this out with black borders, because that will look wrong. That will look like a mistake. People will think we're covering up some of the picture.' I said, 'No they won't, it looks fine. It looks like a movie in a theater.' He goes 'No, no, we should have grey borders. Because that will look more like it's part of the picture and that way they won't think it's a mistake.' I tried to tell him as politely as I could, I think that's the stupidest idea I've ever heard in my life. He goes, 'No, no this is the way we should do it.'"

31. Ross Ruediger, online survey filed at The Video Store Project website, March 6, 2003, available at http://echo.gmu.edu/workshops/jgreenberg/viewdetail.php?responseID=556.

32. David Thomas (video clerk), interview with the author, February 2, 2003.

33. Letterboxing is an arguably controversial practice even as I write this in 2006, and the new 16:9 standard for the shape of high definition television screens promises to keep debates over the shape of televised images in the foreground of public life for years to come.

34. Mimi Kmet, "Hal Roach Studios Puts New Life into Old Black & White Movies," *Video Software Dealer*, October 1985.

35. On early color in film, see Tom Gunning, "Colorful Metaphors: The Attraction of Color in Early Silent Cinema," *Fotogenia* 1, no. 1 (1995); Steve Neale, *Cinema and Technology: Image, Sound, Colour* (London: Macmillan Education Ltd., 1985).

36. Kmet, "Hal Roach Studios Puts New Life into Old Black & White Movies," 28.

37. Ron Horning, "Colorization: Masterstroke or Miscalculation?," *Video Business*, March 1986.

38. Jack Matthews, "Film Directors See Red over Ted Turner's Movie Tinting," *The Los Angeles Times*, September 12, 1986.

39. John Voland, "Turner Defends Move to Colorize Films," *The Los Angeles Times*, October 23, 1986. In addition to his economic claims, Turner's stance arguably reflects a particular understanding of his broadcasts as movies-on-television, subject to the norms and constraints of television, rather than movies in and of themselves. As someone who had made his fortune in broadcasting and was moving into the motion picture industry, it seems logical that Turner would bring his preexisting understandings of the relationship between movies and television into his new broadcasting ventures.

40. Leslie Bennetts, "'Colorizing' Film Classics: A Boon or Bane?" *The New York Times*, August 5, 1986.

41. On the relationship between color television and the movie industry, see Brad Chisholm, "Red, Blue, and Lots of Green: The Impact of Color Television on Feature Film Production," in *Hollywood in the Age of Television*, ed. Tino Balio (Boston: Unwin Hyman, 1990).

42. Eve Allsbrook, "The Coloring of Hollywood," *Video Software Dealer*, February 1987.

43. Colorization also offered the promise of new copyright protections for films that had passed into the public domain, thanks to the decision by the Library of Congress Copyright Office to allow colorized versions of movies to be registered as

copyrightable works, distinct and separate from the copyright status of the original black-and-white film. See Library of Congress Copyright Office, "Copyright Registration for Colorized Versions of Black and White Motion Pictures; Proposed Rulemaking," *Federal Register*, June 24, 1987; Library of Congress Copyright Office, "Notice of Registration Decision," *Federal Register*, June 22, 1987.

44. Jack Matthews, "Sneak Preview of Colorization Wars," *The Los Angeles Times*, October 2, 1986.

45. Bennetts, "'Colorizing' Film Classics: A Boon or Bane?"

46. Tom Huntington, "Black and White in Color," *Saturday Review*, November/December 1985.

47. Ginger Rogers, "Color Me Indignant," *Screen Actor*, Fall 1987.

48. Herb Fischer, "Speaking Out," *Video Software Dealer*, March 1987.

49. R. A. Desbrow, "Let There Be Color," *Los Angeles Herald-Examiner*, October 5, 1986.

50. "CBS Fox Advertisement," *Video Software Dealer*, February 1987.

51 "Pro/Con," *Cable Choice*, February 1987.

52. Vincent Canby, "Film View: 'Colorization' Is Defacing Black and White Film Classics," *The New York Times*, November 2, 1986.

53. Andrew Sarris, "The Color of Money," *The Village Voice*, February 24, 1987.

54. Ted Koppel, *Nightline* (ABC News, November 3, 1986).

55. Horning, "Colorization: Masterstroke or Miscalculation?," 38.

Epilogue

1. Reprinted in Eugene Marlow and Eugene Secunda, *Shifting Time and Space: The Story of Videotape* (Westwood, CT: Praeger Publishers, 1991), 122. For an analysis of this rhetoric of populist power and the VCR, see Kimberly K. Massey and Stanley J. Baran, "VCRs and People's Control of Their Leisure Time" in *Social and Cultura Aspects of VCR Use*, ed. Julia R. Dobrow (Hillsdale, NJ: Lawrence Erlbaum Associates, 1990).

2. Charlyne Varkonyi, "The Truth Behind Those Food Labels: Consumer Groups Say Labeling Is Inadequate," *St. Petersburg Times*, December 29, 1988.

3. According to the reviewer, this joke was one of the highlights of the show. Duncan Strauss, "Comedy Review: Ferrari Proves Loud and Powerful in Irvine but Needs a Tuneup," *Los Angeles Times*, May 5, 1988.

4. Karen S. Peterson, "Skinny-Dippers Count, Too," *USA Today*, May 20, 1991.

5. Varkonyi, "The Truth Behind Those Food Labels."

6. Don Braunagel, "Is Your VCR Smarter Than You? Well, Help Is on the Way," *The San Diego Union-Tribune*, July 13, 1988.

7. "Ted's Quick Fix for VCR Clock: Grab a Washcloth in a Flash," *The Atlanta Journal and Constitution*, May 19, 1991.

8. Greg Beckmann, "How Come They Can Get a Man to the Moon but They Can't Program Their VCRs? Do Rocket Scientists Really Have . . . The Bright Stuff?" *Los Angeles Times*, December 6, 1991.

9. Michael Clancy and Dolores Tropiano, "Don't Feel Bad: Intel VIP's Got a Blinking VCR," *The Arizona Republic*, September 17, 1995.

10. Trudy Drucker, "Progress Meets Its Match," *The New York Times*, April 16, 1989.

11. "Patrolling the Highway," *St. Petersburg Times*, February 15, 1994.

12. David Landis, "Setting Clocks Takes Time," *USA Today*, March 30, 1990.

13. Roberto Peccei and Fred Eiserling, "Literacy for the 21st Century. Education: The New Science Teaching Standards Are Designed to Improve Everybody's Ability," *Los Angeles Times*, February 26, 1996.

14. David Hayes, "Slippery Edge of Digital Age: Technological Marvels Become Commonplace," *The Arizona Republic*, July 26, 1999.

15. Mike Michael, "Ignoring Science: Discourses of Ignorance in the Public Understanding of Science," in *Misunderstanding Science?: The Public Reconstruction of Science and Technology*, ed. Alan Irwin and Brian Wynne (New York: Cambridge University Press, 1996).

16. Marco R. della Cava, "VCR Plus Takes Fuss out of Programming," *USA Today*, November 15, 1990.

17. John Fetto, "Can You Set Your VCR?," *American Demographics* 24, no. 3 (2002).

18. According to some, in fact, it's already dead, though their declarations might be a bit premature (though not terribly so); see Diane Garrett, "VHS, 30, dies of loneliness," *Variety*, November 14, 2005. Available at http://www.variety.com/article/VR1117953955.html.

19. Total annual VCR sales peaked at 27.5 million units in 2000, dropping to 21.8 million in 2001 and 16.9 million in 2002. "Kagan's the State of Home Video 2002," (Carmel, CA: Kagan World Media, 2002), 34–36.

20. David Lieberman, "The Little Tivo That Could: Can It Survive Cable's Attack?," *USA Today*, May 7, 2003.

21. Eugene Kane, "New Way of Watching TV Can Be Life-Altering," *Milwaukee Journal Sentinel*, November 13, 2003.

22. Right now, it looks like TiVo may well find more success from licensing its technology to cable companies than it did as a manufacturer of stand-alone hardware, it being far easier to elaborate on a preexisting technological frame than to create one anew.

23. David Colker, "For Longer-Lasting Memories, Convert Old Tapes to DVDs," *Los Angeles Times*, December 18, 2003. (Ironically, DVDs may well turn out to be a good bit more fragile than magnetic tape as a preservation medium.)

24. David Kaplan, "DVDs Outpace VHS Rentals for First Time," *The Houston Chronicle*, June 20, 2003.

25. Shelley Emling, "Circuit City to Drop VHS Movies," *The Atlanta Journal-Constitution*, June 20, 2002.

26. Peter Dear, "Science Studies as Epistemography," in *The One Culture?: A Conversation About Science*, ed. Jay A. Labinger and Harry Collins (Chicago: University of Chicago Press, 2001).

27. This tension is particularly aptly explored in David Kirby's work on the ways in which professional science consultants delicately manage the identities they present to both scientists and filmmakers. See David Kirby, "Science Consultants, Fictional Films and Scientitic Practice," *Social Studies of Science* 33, no. 2 (2003); David Kirby, "Scientists on the Set: Science Consultants and Communication of Science in Visual Fiction," *Public Understanding of Science* 12 (2003).

28. "Fact Sheet," Netflix.com, http://www.netflix.com/MediaCenter? id=1005 (accessed June 23, 2004).

29. Jim Taylor, "DVD Frequently Asked Questions" DVD Demystified, http://www .dvddemystified.com/dvdfaq.html (accessed June 21, 2004).

30. For an in-depth analysis of this case, see Tarleton Gillespie, "Copyright and Commerce: The DMCA, Trusted Systems, and the Stabilization of Distribution," *The Information Society* 20, no. 4 (2004).

Appendix

1. For a more detailed survey of these initial projects, see Roy Rosenzweig and Dan Cohen, *Digital History* (Philadelphia: University of Pennsylvania Press, 2005), as well as Roy Rosenzweig, *Fostering The Recent History Of Science And Technology In New Media: A Report On Past Practices And Future Directions*, a report presented to the Sloan Foundation, 2000.

2. Exemption 2 of 45 CFR 46.101 states that survey-based research is exempt if it does not "reasonably place the subjects at risk of criminal or civil liability or be damaging to the subjects' financial standing, employability, or reputation."

3. Of 1285 responses, 566, to be precise. A useful experiment would be to randomly offer visitors a survey that is either broken up on multiple pages or entirely on a single page, to test for differences in response rates.

4. Joan Villa, "Back to School." *www.hive4media.com*, March 19, 2003.

5. 'Moriarty,' 'Want to be part of the Video Store Project? Read this to find out!!' *Ain't It Cool News*, http://www.aintitcool.com/display.cgi?id=14636, accessed August 2, 2004.

6. Lauren Cote, online survey conducted on the Video Store Project website, March 6, 2003, available at http://echo.gmu.edu/workshops/jgreenberg/viewdetail.php ?responseID=234.

Index

Inside Technology
edited by Wiebe E. Bijker, W. Bernard Carlson, and Trevor Pinch

.